IET CONTROL, ROBOTICS AND SENSORS SERIES 112

RFID Protocol Design, Optimization, and Security for the Internet of Things

The IET international book series on sensors

IET international book series on sensors – call for authors

The use of sensors has increased dramatically in all industries. They are fundamental in a wide range of applications from communication to monitoring, remote operation, process control, precision and safety, and robotics and automation. These developments have brought new challenges such as demands for robustness and reliability in networks, security in the communications interface and close management of energy consumption. This Book Series covers the research and applications of sensor technologies in the fields of ICTs, security, tracking, detection, monitoring, control and automation, robotics, machine learning, smart technologies, production and manufacturing, photonics, environment, energy and transport.
Book Series Editorial Board

- Dr. Hartmut Brauer, Technische Universität Ilmenau, Germany
- Prof. Nathan Ida, University of Akron, USA
- Prof. Edward Sazonov, University of Alabama, USA
- Prof Desineni "Subbaram" Naidu, University of Minnesota Duluth, USA
- Prof. Wuqiang Yang, University of Manchester, UK
- Prof. Sherali Zeadally, University of Kentucky, USA

Proposals for coherently integrated International multi-authored edited or co-authored handbooks and research monographs will be considered for this Book Series. Each proposal will be reviewed by the IET Book Series editorial board members with additional external reviews from independent reviewers. Please email your book proposal to: vmoliere@theiet.org or author_support@theiet.org.

RFID Protocol Design, Optimization, and Security for the Internet of Things

Alex X. Liu, Muhammad Shahzad, Xiulong Liu and Keqiu Li

The Institution of Engineering and Technology

Published by The Institution of Engineering and Technology, London, United Kingdom

The Institution of Engineering and Technology is registered as a Charity in England & Wales (no. 211014) and Scotland (no. SC038698).

The Institution of Engineering and Technology
Michael Faraday House
Six Hills Way, Stevenage
Herts, SG1 2AY, United Kingdom

www.theiet.org

British Library Cataloguing in Publication Data
A catalogue record for this product is available from the British Library

ISBN 978-1-78561-332-6 (hardback)
ISBN 978-1-78561-333-3 (PDF)

Typeset in India by MPS Ltd

Contents

List of figures

List of tables

Authors' biographies

Alex X. Liu is a Professor of the Department of Computer Science and Engineering at Michigan State University, East Lansing, Michigan, USA. He received his Ph.D. degree in Computer Science from The University of Texas at Austin in 2006. He received the IEEE & IFIP William C. Carter Award in 2004, a National Science Foundation CAREER award in 2009 and the Michigan State University Withrow Distinguished Scholar Award in 2011. He is an Associate Editor of IEEE/ACM Transactions on Networking and IEEE Transactions on Dependable and Secure Computing. He is also an Area Editor of Computer Communications. He received Best Paper Awards from ICNP-2012, SRDS-2012, and LISA-2010. His research interests focus on networking and security.

Muhammad Shahzad is an Assistant Professor in the Department of Computer Science and a member of the Networking Research Group at North Carolina State University, USA. He received his Ph.D. in 2015 from Michigan State University. His research interests is in the broad areas of Networking, Internet of Things and Security, with a focus on RFID based systems, network measurements and modelling, activity recognition and user/device authentication.

Xiulong Liu is a Postdoctoral Fellow in the Department of Computing, The Hong Kong Polytechnic University, China. He received his Ph.D. degree in 2016 from the School of Computer Science and Technology, Dalian University of Technology, China. His research interests focus on RFID Technology and Wireless Networks.

Keqiu Li is a Professor and the Dean of the School of Computer Science and Technology at Tianjin University, China. He got his Ph.D. degree in 2005 from the Japan Advanced Institute of Science and Technology. His research interests focus on data centre networks, cloud computing, multimedia application and wireless network.

Chapter 1

RFID identification—design and optimization

1.1 Introduction

1.1.1 Background and problem statement

As the cost of commercial radio frequency identification (RFID) tags, which is as low as 5 cents per tag [24], has become negligible compared to the prices of the products to which they are attached, RFID systems are being increasingly used in various applications such as supply chain management [15], indoor localization [20], 3D positioning [29], object tracking [19], inventory control, electronic toll collection, and access control [8,18]. For example, Walmart has started to use RFID tags to track jeans and underwear for better inventory control. Large warehouses, such as those of Amazon with sizes up to 1 million ft^2 [2], or distribution centers with sizes up to 3 million ft^2 [1], contain hundreds of thousands of items. RFID systems can make the inventory management and tracking in these large warehouses and distribution centers much easier and error free. An RFID system consists of tags and readers. A tag is a microchip combined with an antenna in a compact package that has limited computing power and communication range. There are two types of tags: (1) passive tags, which do not have their own power source, are powered up by harvesting the radio frequency energy from readers, and have communication ranges often less than 20 ft; (2) active tags, which come with their own power sources and have relatively longer communication ranges. A reader has a dedicated power source with significant computing power. RFID systems mostly work in a query–response fashion where a reader transmits queries to a set of tags and the tags respond with their IDs over a shared wireless medium.

This chapter addresses the fundamental RFID *tag identification* problem, namely reading all IDs of a given set of tags, which is needed in almost all RFID systems. Because tags respond over a shared wireless medium, tag identification protocols are also called *collision arbitration, tag singulation,* or *tag anticollision* protocols. Tag identification protocols need to be scalable as the number of tags that need to be identified could be as large as tens of thousands with the increasing adoption of RFID tags. An RFID system with a large number of tags may require multiple readers with overlapping regions. In this chapter, we first focus on the *single-reader* version of the tag identification problem and then extend our solution to the *multiple-reader* problem.

1.1.2   Summary and limitations of prior art

The industrial standard, EPCGlobal Class 1 Generation 2 (C1G2) RFID [53], adopted two tag identification protocols, namely framed-slotted Aloha and tree walking (TW). In framed-slotted Aloha, a reader first broadcasts a value f to the tags in its vicinity where f represents the number of time slots present in a forthcoming frame. Then each tag whose inventory bit is 0 randomly picks a time slot in the frame and replies during that slot. Each C1G2 compliant tag has an inventory bit, which is initialized to be 0. In any slot, if exactly one tag responds, the reader successfully gets the ID of that tag and issues a command to the tag to change its inventory bit to 1. The key limitation of framed-slotted Aloha is that it cannot identify large tag populations due to the finite possible size of f. Qian *et al.* have shown that framed-slotted Aloha is most efficient when f is equal to the number of tags [22]. Therefore, although theoretically any arbitrarily large tag population can be identified by indefinitely increasing the frame size, practically this is infeasible because during the entire identification process, Aloha-based protocols require all tags, including those that have been identified, to stay powered up and listen to all the messages from the reader in order to maintain the value of the inventory bit. This results in high instability because any intermittent loss of power at a tag will set its inventory bit back to 0, leading the tag to contend in the subsequent frame. The instability of Aloha-based protocols has formally been proven by Rosenkrantz and Towsley in [25].

TW is a fundamental multiple access protocol, which was first invented by U.S. Army for testing soldiers for syphilis during World War II [6]. TW was proposed as an RFID tag identification protocol by Law *et al.* in [14]. In TW, a reader first queries 0 and all the tags whose IDs start with 0 respond. If result of the query is a successful read (i.e., exactly one tag responds) or an empty read (i.e., no tag responds), the reader queries 1 and all the tags whose IDs start with 1 respond. If the result of the query is a collision, the reader generates two new query strings by appending a 0 and a 1 at the *end* of the previous query string and queries the tags with these new query strings. All the tags whose IDs start with the new query string respond. This process continues until all the tags have been identified. This identification process is essentially a partial depth first traversal (DFT) on the complete binary tree over the tag ID space, and the actual traversal forms a binary tree where the leaf nodes represent successful or empty reads and the internal nodes represent collisions. Nodes on level l correspond to lth most significant bit (MSB) of the tag IDs. Figure 1.1(a) shows the TW process for identifying 9 tags over a tag ID space of size 2^4. Here a `successful read` node is one that an identification protocol visits and there is exactly one tag in the subtree rooted at this node, an `empty read` node is one that an identification protocol visits and there is no tag in the subtree rooted at this node, and a `collision` node is one that an identification protocol visits and there are more than one tags in the subtree rooted at this node. The key limitation of TW-based protocols is that they visit a large number of collision nodes in the binary tree, which makes the identification process slow. Although several heuristics have been proposed to reduce the number of visits to collision nodes [17,21], all these

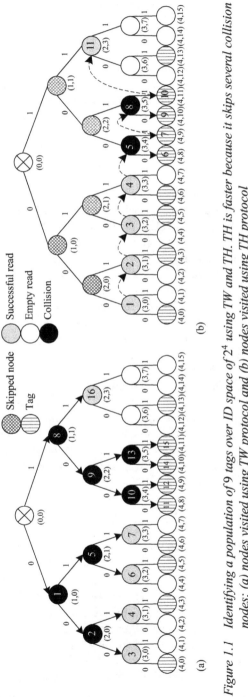

Figure 1.1 Identifying a population of 9 tags over ID space of 2^4 using TW and TH. TH is faster because it skips several collision nodes: (a) nodes visited using TW protocol and (b) nodes visited using TH protocol

heuristics-based methods are not guaranteed to minimize such futile visits. Prior Aloha–TW hybrid protocols also have this limitation.

1.1.3   System model

As most commercially available tags and readers already comply with the C1G2 standard, we do not assume changes to either tags or their physical protocol. We assume that readers can be reprogrammed to adopt new tag identification software. For reliable tag identification, we are given the probability of successful query–response communication between the reader and a tag.

1.1.4   Proposed approach

To address the fundamental limitations that lie in the heuristic nature of prior TW-based protocols, we propose a new approach to tag identification called Tree Hopping (TH). The key novel idea of TH is to formulate the tag identification problem as an optimization problem and find the optimal solution that ensures either minimal expected number of queries (i.e., nodes visited on the binary tree) or minimal expected identification time, as per the requirement. In TH, we first quickly estimate the tag population size. Second, based on the estimated tag population size, we calculate the optimal level to start tree traversal so that the expected number of queries or expected identification time is minimal, *hop* directly to the left most node on that level, and then perform DFT on the subtree rooted at that node. Third, after that subtree is traversed, we reestimate the size of remaining unidentified tag population, recalculate the new optimal level, hop directly to the new optimal node, and perform DFT on the subtree rooted at that node. Hopping to optimal nodes in this manner skips a large number of collision nodes. This process continues until all the tags have been identified. Figure 1.1(b) shows the nodes traversed by TH for the same population of 9 tags as in Figure 1.1(a). Here a `skipped node` is one that TW visits but TH does not. We can see that TH traverses 11 nodes to identify these 9 tags. In comparison, TW traverses 16 nodes as shown in Figure 1.1(a). This difference scales significantly as tag population size increases.

1.1.4.1   Population size estimation

TH first uses a framed-slotted Aloha based method to quickly estimate the tag population size. For this, TH requires each tag to respond to the reader with a probability q. As C1G2 compliant tags do not support this probabilistic responding, we implement this by "virtually" extending the frame size $\frac{1}{q}$ times. To estimate the tag population size, the reader announces a frame size of $\frac{1}{q}$ but terminates it after the first slot. To terminate a frame, the reader issues a SELECT command, specified in the C1G2 standard, with its *position*, *target*, and *action* parameters set to 0. This command "resets" all tags and they go into a state where they expect a new frame to start. For further details on frame termination, see Section 6 of [53]. The reader issues several single-slot frames while reducing q with a geometric distribution (i.e.,, $q = \frac{1}{2^{i-1}}$ in ith frame) until the reader gets an empty slot. Suppose the empty slot occurred in the ith frame, TH estimates the tag population size to be $1.2897 \times 2^{i-2}$ based on Flajolet and Martin's algorithm used in databases [9,23].

1.1.4.2 Finding optimal level

To determine the optimal level γ_{op} that TH directly hops to, we first calculate the expected number of nodes that TH will visit or expected identification time that TH will take if it starts DFTs from nodes on any given level γ. Let b be the number of bits in each tag ID (which is 64 for C1G2 compliant tags), then, we have $1 \leq \gamma \leq b$. If γ is small, more collision nodes will be visited while if it is large, more empty read nodes will be visited. Our objective is to calculate an optimal level γ_{op} that will, depending on the requirement, result in either the smallest number of nodes visited or the smallest identification time. To find γ_{op} for minimizing number of queries, we first derive the expression for calculating the expected number of nodes visited by TH if TH directly hops to level γ. Then, we calculate the value of γ which minimizes this expression. This value of γ is the value of optimal level γ_{op}. We present the technical details of finding γ_{op} in Section 1.3. In Section 1.4, we derive the expression for calculating the expected identification time of TH if TH directly hops to level γ. We use this expression to calculate γ_{op} when we need to minimize the identification time instead of number of queries.

1.1.4.3 Population size reestimation

If the tags that we want to identify are uniformly distributed in the ID space $[0, 2^b - 1]$, then performing DFTs from each node on level γ_{op} will result in minimum number of nodes visited. However, in reality, the tags may not be uniformly distributed. In such cases, each time when the DFT of a subtree is finished, TH needs to reestimate the total tag population size to find the next optimal level and the hoping destination node. TH performs the reestimation as follows. Let z be the first tag population size estimated using the Aloha-based method, x be the number of tags that have been identified, and s be the size of the tag ID space covered by the nodes visited. Naturally, $z - x$ is an estimate of the remaining tag population size; however, we cannot use this estimate to calculate the next optimal level because the remaining leftover ID space may not form a complete binary tree. Instead, based on the node density in the remaining ID space, TH extrapolates the total tag population size to be $\frac{z-x}{2^b-s} \times 2^b$ and uses it to find the next hopping destination node. Note that if tags are uniformly distributed, we have $\frac{z-x}{2^b-s} \times 2^b = z$.

1.1.4.4 Finding hopping destination

Each time after a DFT is done and the new optimal level is recalculated, TH needs to find the next node to hop to, which may not be the leftmost node on the optimal level. Consider the example shown in Figure 1.1(b). Assuming a uniform distribution, the optimal level to start the DFT is 3. In this chapter, we use (l, p) to denote the pth node on level l. TH performs DFTs on the subtrees of nodes $(3, 0)$ to $(3, 5)$ and identifies 8 out of 9 tags. Based on the number of remaining tags after the last DFT, which is 1, the optimal level for the next hop is changed from 3 to 1. However, if TH starts the DFT from the leftmost node on level 1, which is $(1, 0)$, it will result in identifying all tags in its subtree again which is wasteful. Similarly, if TH starts the DFT from the second leftmost node on level 1, which is $(1, 1)$, it will visit the subtree of $(2, 2)$, which is wasteful as all the tags in the subtree of $(2, 2)$ have already been identified. Similarly, if there had been a third leftmost node on the new optimal level and if TH starts the DFT from that third leftmost node, it will not visit the subtree of $(2, 3)$, resulting in tag $(4, 13)$ not being identified. To avoid both scenarios, i.e., some subtrees being traversed multiple times and some subtrees with tags not being traversed, after the optimal level is recalculated, TH hops to the root of the largest subtree that can contain the next tag to be identified but does not contain any previously identified tag. The level at which this root is located can not be smaller than the new optimal level. For the example in Figure 1.1(b), after the subtree rooted at node $(2, 2)$ has been traversed, the recalculated optimal level is 1 and the next node that TH hops to is $(2, 3)$.

Our experimental results in Figure 1.2 show that when the tags are not uniformly distributed in the ID space, our technique of dynamically adjusting γ_{op} according to the leftover population size significantly reduces the total number of queries and the average number of responses per tag. The two curves "TH w reestimation-Seq" and "TH w/o reestimation-Seq" show the total number of queries needed, respectively, with and without the dynamic adjustment of γ_{op} for non-uniformly distributed tag IDs. For example, for 10k tags, this dynamic level adjustment reduces the total number of queries by 31.5%. Our experimental results in Figure 1.2 also show that when the tags are uniformly distributed in the ID space, there is no need to dynamically adjust γ_{op}. The two curves "TH w reestimation-Uni" and "TH w/o reestimation-Uni" show the total number of queries needed, respectively, with and without the dynamic adjustment for uniformly distributed tag IDs. These two curves are similar because for uniformly distributed tag IDs, γ_{op} does not usually change after each DFT and thus the benefit of dynamically adjusting γ_{op} is relatively small. Our experimental results in Figure 1.2 further show that the performance of TH on nonuniformly distributed populations is asymptotically the same as its performance on uniformly distributed

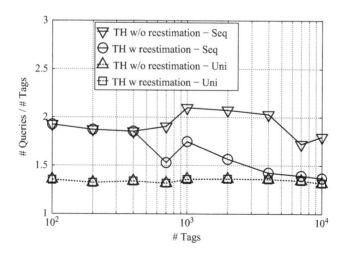

Figure 1.2   Impact of dynamic adjustment of γ_{op} on different types of populations

populations when it uses the technique of dynamically adjusting γ_{op} according to the leftover population size. The curve "TH w reestimation-Seq" approaches the curves "TH w reestimation-Uni" and "TH w/o reestimation-Uni" as the tag population size increases.

1.1.4.5 Population distribution conversion

Although dynamically adjusting γ_{op} for a nonuniformly distributed population reduces the number of queries, the number of queries is still not as low as it would have been had the population been uniformly distributed. Furthermore, the extent of reduction depends on the distribution and size of the population. In Section 1.5.1, we present a simple technique that TH uses to virtually convert almost any nonuniformly distributed population into near-uniformly distributed population. The key idea is that instead of comparing the query strings, transmitted by the reader, with the starting bits of the tag ID, each tag compares the query string with the ending bits of its ID. The resulting binary tree has all the tags near-uniformly distributed in the ID space. We will show that this can be implemented without any modifications to the physical communication protocol and the tags. This technique, combined with the dynamic level adjustment, enables TH to identify any nonuniformly distributed population in almost the same number of queries or time as for uniformly distributed population of the same size. In what follows, we first assume that tags compare the query string with the starting bits of its ID, as in TW protocol, until Section 1.5.1 where we explain this technique in detail.

1.2 Related work

We review existing identification protocols, which can be classified as nondeterministic, deterministic, or hybrid.

1.2.1 Nondeterministic identification protocols

Existing such protocols are either based on framed-slotted Aloha [30] or Binary Splitting (BS) [4]. As we discussed above, Aloha-based protocols only work for small tag populations. In BS [4], the identification process starts with the reader asking the tags to respond. If more than one tag responds, BS divides and subdivides the population into smaller groups until each group has only one or no tag. This process of random subdivision incurs a lot of collisions. Furthermore, BS requires the tags to perform operations that are not supported by the C1G2 standard. ABS is a BS-based protocol that is designed for continuous identification of tags [16].

1.2.2 Deterministic identification protocols

There are 3 such protocols: (1) the basic TW protocol [14], (2) the adaptive TW (ATW) protocol [26], and (3) the TW-based Smart Trend Traversal (STT) protocol [21]. ATW is an optimized version of TW that always starts DFTs from the level of $\log z$, where z is the size of tag population. This is the traditional wisdom for optimizing TW. The key limitation of ATW is that it is optimal only when all tag IDs are evenly spaced in the ID space; however, this is often not true in real-world applications. In contrast, during the identification process, our TH protocol adaptively chooses the optimal level to hop to based on the distribution of IDs. STT improves TW using some ad-hoc heuristics to select prefixes for next queries based upon the type of response to previous queries. It assumes that the number of tags identified in the past k queries is the same as the number of tags that will be identified in the next k queries. This may not be true in reality.

1.2.3   Hybrid identification protocols

Hybrid protocols combine features from nondeterministic and deterministic protocols. There are two major such protocols: Multislotted scheme with Assigned Slots (MAS) [17] and Adaptively Splitting-based Arbitration Protocol (ASAP) [22]. MAS is a TW-based protocol in which each tag that matches the reader's query picks up one of the f time slots to respond. For large populations, due to the finite practical size of f, for queries corresponding to higher levels in the binary tree, the response in each of the f slots is most likely a collision, which increases the identification time. ASAP divides and subdivides the tag population until the size of each subset is below a certain threshold and then applies Aloha on each subset. For this, ASAP requires tags to pick slots using a geometric distribution, which makes it incompliant with the C1G2 standard. Furthermore, subdividing the population before identification is in itself very time consuming.

1.3   Optimal tree hopping

After quick population size estimation using Flajolet and Martin's algorithm [9], TH needs to find the optimal level to hop to. First, we derive an expression to calculate the expected number of queries (i.e., the number of nodes that TH will visit) if it starts DFTs from the nodes on level γ, assuming that tags are uniformly distributed in the ID space. The expression to calculate the expected identification time will be derived in Section 1.4. Second, as the derived expression is too complex to calculate the optimal value of γ that minimizes the expected number of queries by simply differentiating the expression with respect to γ, we present a numerical method to calculate the optimal level γ_{op}. If tags are not uniformly distributed, each time when the DFT on a node is completed, as stated in Section 1.1.4, TH reestimates the total population size based on the initial estimate and the number of tags that have been identified, recalculates the new optimal level, and finds the hopping destination node. Table 1.1 summarizes the symbols used in this chapter.

1.3.1   Average number of queries

Let random variable Q denote the total number of nodes that TH visits to identify all tags. Note that each node visit corresponds to one reader query. We next

Table 1.1   *Notations used in the chapter*

Symbol	Description
b	# Of bits in tag ID, which is 64 for C1G2 tags
n	Size of the whole ID space, which is 2^b
(l,p)	Node whose top-to-down vertical level is $1 \le l \le b$ and left-to-right horizontal position is $0 \le p \le 2^b - 1$
z	Estimated number of tags in the population
γ	Level from which TH performs DFTs
γ_{op}	Optimal level to perform DFTs
q	Tag response probability used in estimation
$I(l,p)$	Indicator random variable. 1 if (l,p) is visited
$I_x(l,p)$	Indicator random variable. 1 if (l,p) is visited and response type is x, where $x \in \{s, c, e\}$
Q	Random variable for total # of nodes visited
$Q_s, Q_c,$ Q_e	Random variables for # of queries resulting in successful reads, collisions, and empty reads
T	Random variable for total identification time
$E[T]$	Expected # of nodes visited to identify all tags in the population
t_s, t_c, t_e	Time duration of successful read, collision, and empty read
$P_l\{(l,p)\}$	Probability of visiting (l,p) if it is left child
$P_r\{(l,p)\}$	Probability of visiting (l,p) if it is right child
m	Size of ID space covered by the parent of the current node being visited
$\overline{m}$	Size of ID space covered by the current node
k	# Of tags covered by the parent of the current node being visited
$\overline{k}$	# Of tags covered by the current node
$P_s, P_c,$ P_e	Probabilities of successful read, collision, or empty read at parent of the current node
$\overline{P}\{x|(l,p)\}$	Probability that response type of node (l,p) is $x \in \{s, c, e\}$ given that node (l,p) is visited
β	Repetitions of query to handle unreliable channel
g	Actual probability of reading a tag
u	Required probability of reading a tag
V	Maximum # of nodes visited to identify all tags
θ_0, θ_1	# Of subtrees covering 0 tags, 1 tags, respectively

calculate $E[Q]$. Let $I(l,p)$ be an indicator random variable whose value is 1 if and only if node (l,p) is visited. Thus, Q is the sum of $I(l,p)$ for all l and all p.

$$Q = \sum_{l=1}^{b} \sum_{p=0}^{2^l-1} I(l,p) \tag{1.1}$$

Let $P\{(l,p)\}$ be the probability that TH visits node (l,p). Thus, $E[Q]$ can be expressed as follows:

$$E[Q] = \sum_{l=1}^{b} \sum_{p=0}^{2^l-1} P\{(l,p)\} \tag{1.2}$$

Next, we focus on expressing $P\{(l,p)\}$ using variable γ, where γ denotes the level that TH hops to. Recall that TH skips all nodes on levels from 1 to $\gamma - 1$ and performs DFT on each of the 2^{γ} nodes on level γ, where $1 \leq \gamma \leq b$. Note that the root node of the whole binary tree is always meaningless to visit as it corresponds to a query of length 0. Here $P\{(l,p)\}$ is calculated differently depending on whether node (l,p) is the left child of its parent or the right. Let $P_l\{(l,p)\}$ and $P_r\{(l,p)\}$ denote the probability of visiting (l,p) when (l,p) is the left and right child of its parent, respectively. If the estimated total number of tags z is zero, then $P_l\{(l,p)\} = P_r\{(l,p)\} = 0$ for all l and p. Below we assume $z > 0$. As TH skips all nodes from levels 1 to $\gamma - 1$, we have

$$P_l\{(l,p)\} = P_r\{(l,p)\} = 0 \text{ if } 1 \leq l < \gamma \tag{1.3}$$

As TH performs DFT from each node on level γ, it visits each node on this level. Thus, we have

$$P_l\{(l,p)\} = P_r\{(l,p)\} = 1 \text{ if } l = \gamma \tag{1.4}$$

For each remaining level $\gamma < l \leq b$, when (l,p) is the left child of its parent, $P_l\{(l,p)\}$ is equal to the probability that the parent of (l,p) is a collision node. When (l,p) is the right child of its parent, if the parent is a collision node and $(l,p-1)$ is an empty read node, then (l,p) will also be a collision node. Thus, instead of visiting (l,p), TH should directly hop to the left child of (l,p). Therefore, $P_r\{(l,p)\}$ is equal to the probability that the parent of (l,p) is a collision node and $(l,p-1)$ is not an empty read node.

Let k denote the number of tags *covered* by the parent of node (l,p) [i.e., the number of tags that are in the subtree rooted at the parent of (l,p)]. Let $m = 2^{b-l+1}$ denote the maximum number of tags that the parent of (l,p) can cover and $n = 2^b$ denote the maximum number of tags that can be accommodated in the whole ID space. The probability that the parent of (l,p) covers k of z tags follows a hypergeometric distribution:

$$P\{\#\text{tags} = k\} = \frac{\binom{m}{k}\binom{n-m}{z-k}}{\binom{n}{z}} \tag{1.5}$$

Let P_e be the probability that the parent of (l,p) is an empty read. Thus,

$$P_e = P\{\#\text{tags} = 0\} = \frac{\binom{n-m}{z}}{\binom{n}{z}} \tag{1.6}$$

Let P_s be the probability that the parent of (l,p) is a successful read. Thus,

$$P_s = P\{\#\text{tags} = 1\} = \frac{m\binom{n-m}{z-1}}{\binom{n}{z}} \tag{1.7}$$

Let P_c be the probability that the parent of (l,p) is a collision node. Thus,

$$P_c = 1 - (P_e + P_s) = 1 - \frac{\binom{n-m}{z}}{\binom{n}{z}} - \frac{m\binom{n-m}{z}}{\binom{n}{z}} \tag{1.8}$$

Next we calculate $P_l\{(l,p)\}$ and $P_r\{(l,p)\}$ for $\gamma < l \le b$ for the following three cases: $n - m < z - 1$, $n - m = z - 1$, and $n - m > z - 1$. Note that $n - m$ is the size of the ID space that is not covered by the parent of (l,p), and $z - k$ is the remaining number of tags that are not covered by the parent of (l,p). Thus, $z - k \le n - m$.

Case 1

$n - m < z - 1$. In this case, $z - k \le n - m < z - 1$, which means $k \ge 2$. Thus, as the parent of (l,p) covers at least two tags, it must be a collision node, i.e., $P_c = 1$. Thus, if (l,p) is the left child of its parent, TH for sure visits it:

$$P_l\{(l,p)\} = 1 \tag{1.9}$$

If (l,p) is the right child of its parent, TH visits it if and only if node $(l,p-1)$, which is the left sibling of (l,p), is not an empty read. If $(l,p-1)$ is an empty read, as its parent is a collision node, (l,p) must also be a collision node, which means that TH will directly visit the left child of (l,p) instead of (l,p). The size of the ID space covered by $(l,p-1)$ is $\frac{m}{2}$. If $n - \frac{m}{2} \le z - 1$, then node $(l,p-1)$ covers at least one tag, which means that $(l,p-1)$ is not an empty read and TH for sure visits (l,p), i.e., $P_r\{(l,p)\} = 1$. If $n - \frac{m}{2} > z - 1$, then the probability that TH visits (l,p) is equal to the probability that $(l,p-1)$ is not an empty read, which is $1 - \binom{n-\frac{m}{2}}{z}/\binom{n}{z}$ based on (1.6). Finally, we have

$$P_r\{(l,p)\} = \begin{cases} 1 - \frac{\binom{n-\frac{m}{2}}{z}}{\binom{n}{z}} & \text{if } n - \frac{m}{2} > z - 1 \\ 1 & \text{if } n - \frac{m}{2} \le z - 1 \end{cases} \tag{1.10}$$

Case 2

$n - m = z - 1$. In this case, $z - k \le n - m = z - 1$, which means $k \ge 1$. As the parent of (l,p) covers $k \ge 1$ tags, the probability of the parent of (l,p) being an empty read is 0 and the probability of the parent of (l,p) being a successful read is $m\binom{z-1}{z-1}/\binom{n}{z} = m\binom{z-1}{z-1}/\binom{n}{z} = m/\binom{n}{z}$ based on (1.7). If (l,p) is the left child of its parent, then TH visits it if and only if the parent of (l,p) is a collision node. Thus, the probability of visiting (l,p) is equal to the probability of the parent of (l,p) being a collision node, which is equal to $1 - P_e - P_s$. Thus, we have

$$P_l\{(l,p)\} = 1 - P_e - P_s = 1 - \frac{m}{\binom{n}{z}} \tag{1.11}$$

If (l,p) is the right child of its parent, then TH visits it if and only if both the parent of (l,p) is a collision node and $(l,p-1)$ is not an empty read. The probability that the parent of (l,p) is a collision node is $1 - m/\binom{n}{z}$ as calculated above. Given that the parent of (l,p) is a collision node, the probability that $(l,p-1)$ is an empty read is $\left(\binom{n-\frac{m}{2}}{z} - \frac{m}{2} \right) / \left(\binom{n}{z} - m \right)$.

$$P_r\{(l,p)\} = \left[1 - \frac{m}{\binom{n}{z}} \right] \cdot \left[1 - \frac{\binom{n-\frac{m}{2}}{z} - \frac{m}{2}}{\binom{n}{z} - m} \right] \tag{1.12}$$

Case 3

$n - m > z - 1$. In this case, $k \geq 0$. Similar to the calculations above, as per (1.6) and (1.7), we have:

$$P_l\{(l,p)\} = 1 - P_e - P_s = 1 - \frac{\binom{n-m}{z} + m\binom{n-m}{z-1}}{\binom{n}{z}} \tag{1.13}$$

$$P_r\{(l,p)\} = \left[1 - \frac{\binom{n-m}{z} + m\binom{n-m}{z-1}}{\binom{n}{z}} \right]$$
$$\times \left[1 - \frac{\binom{n-\frac{m}{2}}{z} - \left\{ \binom{n-m}{z} + \frac{m}{2}\binom{n-m}{z-1} \right\}}{\binom{n}{z} - \left\{ \binom{n-m}{z} + m\binom{n-m}{z-1} \right\}} \right] \tag{1.14}$$

Finally, (1.3)–(1.14) completely define the probabilities $P_l\{(l,p)\}$ and $P_r\{(l,p)\}$. Note that as tags are uniformly distributed, the probability of visiting node (l,p) is independent of the horizontal position p.

The expected number of queries can now be calculated using Theorem 1.1.

Theorem 1.1. *For a population of z tags uniformly distributed in the ID space, where each tag has an ID of b bits, if TH hops to level γ to perform DFT from each node on this level, the expected number of queries for identifying all z tags is:*

$$E[Q] = 2^\gamma + \sum_{l=\gamma+1}^{b} 2^{l-1}[P_l\{(l,p)\} + P_r\{(l,p)\}] \tag{1.15}$$

Proof. First, on level γ, all the 2^γ nodes are visited by TH. Second, on any level l where $\gamma + 1 \leq l \leq b$, the probabilities of left and right nodes being visited are $P_l\{(l,p)\}$ and $P_r\{(l,p)\}$, respectively. As there are 2^{l-1} pairs of left and right nodes on level l, the expected number of nodes visited by TH on level l is $2^{l-1}[P_l\{(l,p)\} + P_r\{(l,p)\}]$. □

When $\gamma = 1$, (1.15) is also the analytical model for calculating expected number of queries of TW protocol.

1.3.2 Calculating optimal hopping level

Equation (1.15) shows that $E[Q]$ is a function of γ as $n = 2^b$, $m = 2^{b-l+1}$, and b is given. For any given z, we want to find the optimal level $\gamma = \gamma_{op}$ so that $E[Q]$ is minimal. The conventional approach to finding the optimal variable value that minimizes a given function is to differentiate the function with respect to that variable, equate the resulting expression to zero, and solve the equation to obtain the optimal

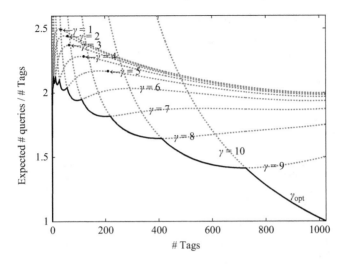

Figure 1.3   Norm. $E[Q]$ vs. population size $\forall \gamma$

variable value. However, it is very difficult, if not impossible, to use this approach to find the optimal level because (1.15) for calculating $E[Q]$ is too complex.

Next, we present a numerical method to find the optimal level. First, we define *normalized $E[Q]$* as the ratio of $E[Q]$ to tag population size. Figure 1.3 shows the plots of normalized $E[Q]$ vs. the number of tags for different γ values ranging from 1 to b (here we used $b = 10$ for illustration). From this figure, we observe that for any tag population size, there is a unique optimal value of γ. For example, for a population of 600 tags, $\gamma_{op} = 9$. Second, we define *crossover points* as follows: *for a given ID length b, the crossover points are the tag population sizes $c_0 = 0, c_1, c_2, \ldots, c_{b+1} = 2^b$ such that for any tag population size in $[c_i, c_{i+1})$ $(0 \le i \le b)$, $\gamma_{op} = i$.* These crossover points are essentially the x-coordinates of the intersection points of the normalized $E[Q]$ curves of consecutive values of γ in Figure 1.3. Thus, the value of c_i can be obtained by putting $z = c_i$ and numerically solving $E[Q, \gamma = i - 1] = E[Q, \gamma = i]$ for c_i using the bisection method. Once c_i is calculated for each $1 \le i \le b$, γ_{op} for a given z can be obtained by simply identifying the unique interval $[c_i, c_{i+1})$ in which z lies and then using $\gamma_{op} = i$. The solid line in Figure 1.3 is plotted using the values of γ_{op} obtained using the proposed strategy. As values of c_i only depend on b, it is a one-time cost to calculate them. Table 1.2 tabulates the values of c_i obtained using this strategy for $b = 64$.

We next conduct an analytical comparison between the expected number of queries for TH and that for TW. Figure 1.4 shows the expected number of queries for TH, which is calculated using (1.15) using $\gamma = \gamma_{op}$, and that for TW, which is calculated using (1.15) using $\gamma = 1$, for 64 bit tag IDs. We observe that TH significantly outperforms TW for the expected number of queries. For example, for a population of 10k tags, the expected number of queries for TH is only 54% of that

Table 1.2   All crossover points for 64-bit tag IDs

c_0	0.00E+00	c_{22}	3.84E+06	c_{44}	1.61E+13
c_1	2.00E+00	c_{23}	7.68E+06	c_{45}	3.22E+13
c_2	4.00E+00	c_{24}	1.54E+07	c_{46}	6.45E+13
c_3	7.00E+00	c_{25}	3.07E+07	c_{47}	1.29E+14
c_4	1.50E+01	c_{26}	6.12E+07	c_{48}	2.58E+14
c_5	2.90E+01	c_{27}	1.22E+08	c_{49}	5.16E+14
c_6	5.90E+01	c_{28}	2.42E+08	c_{50}	1.03E+15
c_7	1.17E+02	c_{29}	4.76E+08	c_{51}	2.06E+15
c_8	2.35E+02	c_{30}	9.22E+08	c_{52}	4.13E+15
c_9	4.69E+02	c_{31}	1.73E+09	c_{53}	8.25E+15
c_{10}	9.38E+02	c_{32}	3.04E+09	c_{54}	1.65E+16
c_{11}	1.88E+03	c_{33}	8.59E+09	c_{55}	3.30E+16
c_{12}	3.75E+03	c_{34}	1.72E+10	c_{56}	6.59E+16
c_{13}	7.51E+03	c_{35}	3.01E+10	c_{57}	1.32E+17
c_{14}	1.50E+04	c_{36}	6.44E+10	c_{58}	2.63E+17
c_{15}	3.00E+04	c_{37}	1.25E+11	c_{59}	5.24E+17
c_{16}	6.00E+04	c_{38}	2.53E+11	c_{60}	1.04E+18
c_{17}	1.20E+05	c_{39}	5.03E+11	c_{61}	2.04E+18
c_{18}	2.40E+05	c_{40}	1.01E+12	c_{62}	3.96E+18
c_{19}	4.80E+05	c_{41}	2.01E+12	c_{63}	7.42E+18
c_{20}	9.61E+05	c_{42}	4.03E+12	c_{64}	1.30E+19
c_{21}	1.92E+06	c_{43}	8.06E+12	c_{65}	1.84E+19

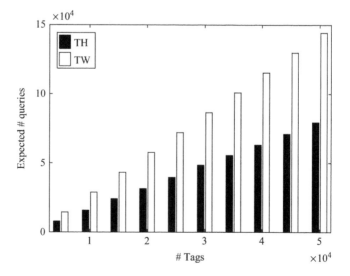

Figure 1.4   E[Q]: TH vs. TW

for TW. We will present detailed experimental comparison between TH and other identification protocols in Section 1.6.

1.3.3 Maximum number of queries

Although the primary goal of our TH protocol is to minimize the average number of queries, next, we analyze the maximum number of queries of TH and analytically show that it is still smaller than that of TW.

The maximum number of queries that TH may need to identify z tags with b-bit IDs is shown in Theorem 1.2.

Theorem 1.2. *Let V denote the number of queries that TH may need to identify a population of $z \geq 2$ tags with b-bit IDs using $\gamma = \gamma_{op}$. We have*

$$V \leq z(b - \gamma_{op} + 1) - 2^{\gamma_{op}} + 2\theta_0 - \theta_1(b - \gamma_{op} - 1) \tag{1.16}$$

where

$$\theta_0 = 2^{\gamma_{op}} - \left\lceil \frac{z}{2^{b-\gamma_{op}}} \right\rceil \tag{1.17}$$

$$\theta_1 = \left\lceil \frac{z}{2^{b-\gamma_{op}}} \right\rceil - \left\lceil \frac{z-1}{2^{b-\gamma_{op}}} \right\rceil \left\lceil 1 - \frac{\gamma_{op}}{b} \right\rceil \tag{1.18}$$

Proof. Let V_{TW} denote the number of queries that TW may need to identify $z \geq 2$ tags with b-bit IDs. The upper bound of V_{TW} is given as follows (proven in [14]):

$$V_{TW} \leq z\left(b + 1 - \log \frac{z}{2}\right) - 1 \tag{1.19}$$

Because $z \geq 2$, we have $V_{TW} \leq z(b+1) - 1$.

When z tags are uniformly distributed in the ID space, TH essentially performs TW on all subtrees rooted at nodes on level γ_{op}. Let θ_0 and θ_1 denote the number of subtrees covering 0 and 1 tags, respectively.

For these $\theta_0 + \theta_1$ subtrees, TH only visits the roots, which are at level γ_{op}. Let α denote the number of remaining subtrees (i.e., $\alpha = 2^{\gamma_{op}} - \theta_0 - \theta_1$) and T_i denote a subtree covering $z_i \geq 2$ tags. For each subtree T_i, the maximum number of nodes that TH visits is $z_i(b - \gamma_{op} + 1) - 1$. Summing all $2^{\gamma_{op}}$ subtrees, we have

$$V \leq \sum_{i=0}^{\alpha-1} \left(z_i(b - \gamma_{op} + 1) - 1\right) + \theta_0 + \theta_1$$

$$= z(b - \gamma_{op} + 1) - 2^{\gamma_{op}} + 2\theta_0 - \theta_1(b - \gamma_{op} - 1) \tag{1.20}$$

The right hand side (RHS) of (1.20) is maximized when θ_0 is maximized and θ_1 is minimized, which happens when all z tag IDs are contiguous and they start from the left most leaf of a subtree at level γ_{op}. In this case, the number of subtrees with tags are $\left\lceil \frac{z}{2^{b-\gamma_{op}}} \right\rceil$ and therefore $\theta_0 = 2^{\gamma_{op}} - \left\lceil \frac{z}{2^{b-\gamma_{op}}} \right\rceil$. Furthermore in this case, when $\gamma_{op} \leq b - 1$, there is at most one subtree at level γ_{op} that has exactly one tag i.e.,

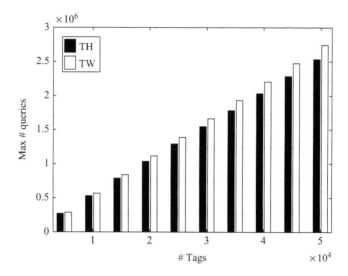

Figure 1.5   Max. # queries: TH vs. TW

$\theta_1 = \left\lceil \frac{z}{2^{b-\gamma_{op}}} \right\rceil - \left\lceil \frac{z-1}{2^{b-\gamma_{op}}} \right\rceil$; when $\gamma_{op} = b$, θ_1 equals z. Combining the two cases of $\gamma_{op} \leq b - 1$ and $\gamma_{op} = b$, we have $\theta_1 = \left\lceil \frac{z}{2^{b-\gamma_{op}}} \right\rceil - \left\lceil \frac{z-1}{2^{b-\gamma_{op}}} \right\rceil \left\lceil 1 - \frac{\gamma_{op}}{b} \right\rceil$.   □

The proof above gives us the insight that *TH requires fewer queries when the tag IDs are distributed more uniformly in the ID space.* Intuitively, this makes sense because the more the tag IDs are distributed uniformly, the fewer the number of collisions encountered by TH. Experimentally, our results shown in Figure 1.11(a) and 1.11(b) in Section 1.6 also confirm this insight: for the same number of tags, the number of queries needed by TH when tags are uniformly distributed is less than that when tags are nonuniformly distributed.

We now conduct an analytical comparison between the maximum number of queries for TH and that for TW. Figure 1.5 shows the maximum number of queries for TH, which is calculated using the RHS of (1.16), and that for TW, which is calculated using the RHS of (1.19), for 64 bit tag IDs. We observe that TH again outperforms TW for the maximum number of queries, although slightly. For example, for a population of 10k tags, the maximum number of queries for TH is 93% of that for TW.

1.4   Minimizing identification time

The optimal value of γ calculated using the expression for $E[Q]$ in (1.15) and applying the numerical method proposed in Section 1.3.2 minimizes the average number of queries, but does not minimize the average identification time because the durations of successful read, empty read, and collision are different. Next, we derive an expression for expected identification time as a function of γ. We can then use the numerical

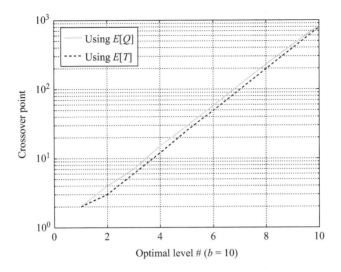

Figure 1.6   *Crossover points obtained using E[Q] and E[T]*

method of Section 1.3.2 to calculate the optimal value of γ that will minimize the average identification time.

Let random variable T denote the total identification time that TH takes to identify all tags. Next, we calculate $E[T]$. Let t_s, t_c, and t_e denote the time durations of successful read, collision, and empty read, respectively. Let random variables Q_s, Q_c, and Q_e denote the number of queries resulting in successful reads, collisions, and empty reads, respectively. Thus, T can be expressed as follows:

$$T = Q_s \times t_s + Q_c \times t_c + Q_e \times t_e \tag{1.21}$$

Applying expectation operator on both sides of the equation above, the expected value of total identification time, $E[T]$, can be expressed as follows:

$$E[T] = E[Q_s] \times t_s + E[Q_c] \times t_c + E[Q_e] \times t_e \tag{1.22}$$

Next, we derive expressions for $E[Q_s]$, $E[Q_c]$, and $E[Q_e]$. Let $I_x(l,p)$ be an indicator random variable whose value is 1 if and only if node (l,p) is visited and the response type is x, where $x \in \{$s:successful read, c:collision, e:empty read$\}$. Thus, Q_x is the sum of $I_x(l,p)$ for all l and all p, where $x \in \{$s, c, e$\}$.

$$Q_x = \sum_{l=1}^{b} \sum_{p=0}^{2^l-1} I_x(l,p) \tag{1.23}$$

The probability that TH visits node (l,p) is $P\{(l,p)\}$. Let $\overline{P}\{x|(l,p)\}$ be the probability that given that TH visits node (l,p), the response type for the node is x, where $x \in \{$s, c, e$\}$. Thus, $E[Q_x]$ can be expressed as follows:

$$E[Q_x] = \sum_{l=1}^{b} \sum_{p=0}^{2^l-1} P\{(l,p)\} \times \overline{P}\{x|(l,p)\} \tag{1.24}$$

Recall that $P\{(l,p)\}$ has already been completely defined in (1.3)–(1.14). Next, we derive expressions for $\overline{P}\{x|(l,p)\}$. Let $\overline{k}$ denotes the number of tags covered by the node (l,p). Let $\overline{m} = 2^{b-l}$ denotes the maximum number of tags node (l,p) can cover. Recall that $n = 2^b$ denotes the maximum number of tags that can be accommodated in the whole ID space. The probability that node (l,p) covers $\overline{k}$ of z tags follows a hypergeometric distribution:

$$P\left\{\#\text{tags} = \overline{k}\right\} = \frac{\binom{\overline{m}}{\overline{k}}\binom{n-\overline{m}}{z-\overline{k}}}{\binom{n}{z}} \tag{1.25}$$

The probabilities $\overline{P}\{s|(l,p)\}$ and $\overline{P}\{e|(l,p)\}$ can be calculated using $\overline{k} = 1$ and $\overline{k} = 0$, respectively, in (1.25).

$$\overline{P}\{s|(l,p)\} = \left\{ \begin{array}{ll} \frac{\overline{m}\binom{n-\overline{m}}{z-1}}{\binom{n}{z}} & \text{if } n - \overline{m} \geq z - 1 \\ 0 & \text{if } n - \overline{m} < z - 1 \end{array} \right\} \tag{1.26}$$

$$\overline{P}\{e|(l,p)\} = \left\{ \begin{array}{ll} \frac{\binom{n-\overline{m}}{z}}{\binom{n}{z}} & \text{if } n - \overline{m} > z - 1 \\ 0 & \text{if } n - \overline{m} \leq z - 1 \end{array} \right\} \tag{1.27}$$

Probability $P\{c|(l,p)\}$ can be calculated as follows:

$$\overline{P}\{c|(l,p)\} = 1 - (\overline{P}\{e|(l,p)\} + \overline{P}\{s|(l,p)\})$$
$$= \left\{ \begin{array}{ll} 1 - \frac{\overline{m}\binom{n-\overline{m}}{z-1}}{\binom{n}{z}} - \frac{\binom{n-\overline{m}}{z}}{\binom{n}{z}} & \text{if } n - \overline{m} > z - 1 \\ 1 - \frac{\overline{m}}{\binom{n}{z}} & \text{if } n - \overline{m} = z - 1 \\ 0 & \text{if } n - \overline{m} < z - 1 \end{array} \right\} \tag{1.28}$$

The expected identification time of TH can now be calculated using Theorem 1.3.

Theorem 1.3. *For a population of z tags uniformly distributed in the ID space, where each tag has an ID of b bits, if TH hops to level γ to perform DFT from each node on this level, the expected identification time for identifying all z tags is*

$$E[T] = 2^{\gamma}[t_c + (t_s - t_c)\overline{P}\{s|(\gamma,p)\} + (t_e - t_c)\overline{P}\{e|(\gamma,p)\}]$$

$$+ \sum_{l=\gamma+1}^{b} \{[t_c + (t_s - t_c)\overline{P}\{s|(l,p)\} + (t_e - t_c)\overline{P}\{e|(l,p)\}] \tag{1.29}$$

$$\times 2^{l-1}[P_l\{(l,p)\} + P_r\{(l,p)\}]\}$$

Proof. Equation (1.29) is obtained in three steps. First, substitute the values of $\overline{P}\{s|(l,p)\}$, $\overline{P}\{e|(l,p)\}$, and $\overline{P}\{c|(l,p)\}$ from (1.26), (1.27), and (1.28) into (1.24) to obtain values of $E[Q_s]$, $E[Q_e]$, and $E[Q_c]$, respectively, and further substitute these values of $E[Q_s]$, $E[Q_e]$, and $E[Q_c]$ into (1.22). Second, use $P\{(l,p)\} = 0$ for $1 \leq l < \gamma$ as per (1.3) and use $P\{(l,p)\} = 1$ for $l = \gamma$ as per (1.4). Third, for any level $l > \gamma$, use $P\{(l,p)\} = P_l\{(l,p)\}$ for each node on this level that is left child of its parent and use $P\{(l,p)\} = P_r\{(l,p)\}$ for each node on this level that is right child of its parent. Note that there are 2^{l-1} pairs of left and right nodes on level l. □

When $\gamma = 1$, (1.29) is also the analytical model for calculating expected identification time of TW protocol. Note that (1.29) is a generalized form of (1.15). It reduces to (1.15) if the time durations of successful read, collision, and empty read are equal to unit time.

According to [11,53], the values of t_s, t_e, and t_c are 3, 0.3, 1.5 ms, respectively. Figure 1.7 plots the values of crossover points obtained using expression of $E[Q]$ from Theorem 1.1 and expression of $E[T]$ from Theorem 1.3 (we used $b = 10$ for illustration). We observe from the figure that the values of crossover points obtained using the expression for $E[Q]$ are comparatively larger than those obtained using the expression for $E[T]$. The reason is that to minimize identification time instead of number of queries, TH starts the DFTs at levels with comparatively larger values of l, which results in reduction in number of collisions at an expense of slightly increased number of empty reads. The overall identification time is reduced because empty reads are five times faster than collisions and the amount of identification time increased by the increased number of empty reads is smaller than the amount of identification time reduced by the reduced number of collisions. Figure 1.8 shows the normalized expected identification times for the two cases i.e., when the crossover points are calculated using $E[Q]$ and $E[T]$ (again we used $b = 10$ for illustration). We observe that for several population sizes, the normalized expected time calculated using $E[Q]$ is greater than that calculated using $E[T]$.

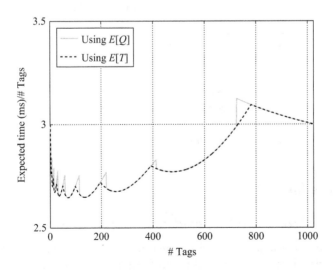

Figure 1.7 Normalized expected identification time vs. population size

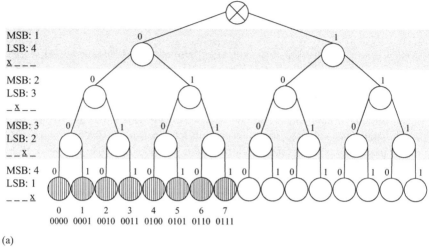

(a)

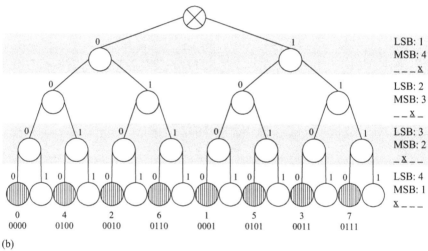

(b)

Figure 1.8   Distributions of populations when binary trees are made with MSBs and LSBs. The population is more uniformly in trees made with LSBs. (a) Binary tree where level l corresponds to lth MSB of the tag ID and (b) binary tree where level l corresponds to lth LSB of the tag ID

1.5   Discussion

1.5.1   Virtual conversion of population distributions

To virtually convert a nonuniformly distributed population into a uniformly distributed population, we leverage the fact that in large populations, the expected number of tags whose IDs have the least significant bit (LSB) of 0 is approximately the same as the expected number of tags whose IDs have the LSB of 1. Similarly, the expected number of tags whose IDs have the two LSBs of 00 is approximately the same as the expected number of tags whose IDs have the two LSBs of 01, 10, or 11, and so on. Therefore, if we construct a binary tree in which level l corresponds to lth LSB instead of lth MSB, then each node of level l is expected to cover $z/2^l$ tags: a property of uniformly distributed populations. To illustrate, consider an example where there are 8 tags in a population, each with a unique 4-bit ID in the range [0, 7]. Figure 1.9(a) shows the binary tree constructed in the conventional way in which level l corresponds to lth MSB. This population is clearly nonuniformly distributed in the ID space and TH will have to frequently perform dynamic adjustments to the optimal value of γ and the number of queries will be large compared to the number of queries for a uniformly distributed population of the same size. Figure 1.9(b) shows the binary tree constructed in the proposed way where level l corresponds to lth LSB. Note from the figure that the 8 tags are now uniformly placed in the entire ID space. On the binary trees that resembles the one in Figure 1.9(b), TH will require very few dynamic adjustments and the number of queries will be approximately same as for a uniformly distributed population of the same size.

Figure 1.10(a1)–1.10(c2) shows three other populations where the circles on the left side of the dashed vertical line represent level $l = b = 4$ of the binary tree in which level l corresponds the lth MSB of the tag ID, and the circles on the right side of the dashed vertical line represent level $l = b = 4$ of the binary tree in which level l corresponds the lth LSB of the tag ID. The population in Figure 1.10(a1) and 1.10(a2) consists of 8 tags with consecutive IDs in the range [4, 11]. We can see that if the binary tree is built using conventional method where lth level corresponds to the lth MSB of the tag ID, then the resulting population is not uniformly distributed in the binary tree. However, if the binary tree is built using our proposed modification where lth level corresponds to the lth LSB of the tag ID, then the resulting population is more close to a uniform distribution. Similarly, the population in Figure 1.10(b1) and 1.10(b2) consists of two blocks, each containing 3 IDs. We make the same observation that the IDs are comparatively more uniformly distributed in the binary tree made with LSBs compared to the one made with MSBs. In a scenario where a population is already uniformly distributed in the ID space, our proposed modification does not affect it and the uniformity is maintained in the tree made with LSBs. This is shown in Figure 1.10(c1) and 1.10(c2).

Next, we leverage these observations to propose a simple modification in TH that reduces the number of queries and identification times of TH for nonuniformly distributed populations to approximately the same values as for uniformly distributed populations. When the reader transmits a query string, the tag compares it with its LSBs instead of MSBs to decide whether or not it will respond to the query. If the result of the query is a collision, the reader generates two new query strings by appending a 0 and a 1 at the *start* of the previous query string and queries the tags with these new query strings. All the tags whose IDs end with the new query string respond.

This modification does not require any changes to the tags and works with the C1G2 compliant tags. To make a tag compare the query string with the LSBs of its ID, we use the SELECT command standardized in the C1G2 standard. The ID of a

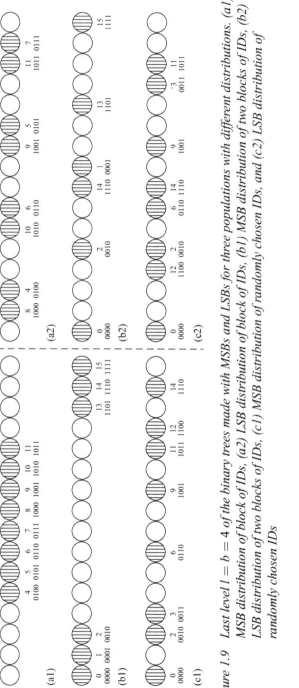

Figure 1.9 Last level l = b = 4 of the binary trees made with MSBs and LSBs for three populations with different distributions. (a1) MSB distribution of block of IDs, (a2) LSB distribution of block of IDs, (b1) MSB distribution of two blocks of IDs, (b2) LSB distribution of two blocks of IDs, (c1) MSB distribution of randomly chosen IDs, and (c2) LSB distribution of randomly chosen IDs

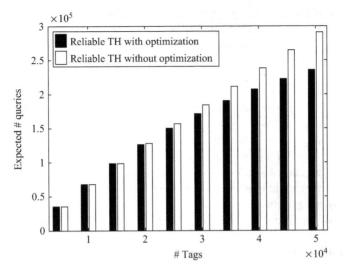

Figure 1.10 E[Q] of Reliable TH

tag is stored in its memory at a specific memory address. A tag can retrieve any bits stored in its memory by specifying an appropriate address range. Using the SELECT command, a reader broadcasts an address range and a bit mask. Each tag compares the bit mask with the bits in the specified address range in its memory and responds back only if the bit mask matches the specified bits in its memory. In TH, the bit mask contains the query string of length l, where $1 \leq l \leq b$, and the address range that the reader broadcasts is of the l LSBs of tag IDs.

1.5.2 Reliable tag identification

So far we have assumed that the communication channel between the reader and tags is reliable, which means that each tag can receive the query from the reader and the reader can receive either the response if only one tag responds or the collision if more than one tag respond. However, this assumption often does not hold in reality because wireless communication medium is inherently unreliable. There are two existing schemes for making tag identification reliable. Backes *et al.* proposed the scheme of letting each tag store the IDs of several other tags [3]. When the reader queries a tag, the tag transmits back its own ID as well as the IDs of other tags stored in it. When identification completes, the reader compares the set of IDs of tags that responded with the union of sets of IDs of other tags reported by each responding tag. If the sets are not equal, the whole process is repeated again to ensure that the missed tags are identified. This scheme has two weaknesses. First, this scheme does not comply with the C1G2 standard. Second, it assumes that the tag population remains static for the lifetime of tags as each tag is hard coded with some other tags' IDs. The second scheme is to run an identification protocol on the same population several times until probability of missing a tag falls below a threshold [10,12]. They estimate the probability of missing a tag based upon the number of tags that were identified in some runs of the protocol but not in others.

While we can use the C1G2 compliant scheme proposed in [10,12] to make TH reliable, i.e., repeatedly run TH until the required reliability is achieved. We observe that in this scheme, the leaf nodes in the binary tree are queried multiple times. This is wasteful of time for the nodes that the reader successfully reads. To eliminate such waste, we propose *to query each node multiple times, instead of querying the whole binary tree multiple times*. We define the *reliability of successfully reading a tag* to be *the probability that both the tag receives the query from the reader and the reader receives the response from the tag*. For this, we calculate the maximum number of times the reader should transmit a query, which is denoted by β. Let g and u be the *given* and *required* reliability of successfully reading a tag, respectively. Thus, the probability of successfully identifying a tag is $1 - (1 - g)^\beta$. Equating it to u, we get:

$$\beta = \log_{(1-g)}(1 - u) \tag{1.30}$$

Our scheme of reliable tag identification works as follows: for each nonterminal node in the binary tree that TH needs to visit, TH transmits a query corresponding to that node β times; corresponding to each terminal node, TH keeps transmitting the query until either that query has been transmitted β times or the reader successfully receives the tag ID.

The optimization technique of stop transmitting the query corresponding to a terminal node on a successful read significantly reduces the total number of queries. Figure 1.6 plots the expected number of queries per tag for the reliable TH protocol with and without this optimization. For example, for a population of 50,000 tags, the number of queries per tag are reduced by 24%.

1.5.3   *Continuous scanning*

In some applications, the tag population may change over time (i.e., tags leave and join the population dynamically). We adapt the continuous scanning strategy proposed by Myung *et al.* in [16]. In the first scanning of the whole tag population, TH records the queries that resulted in successful or empty reads. If the tag population does not change, by performing DFTs on the subtrees rooted at successful and empty read nodes of the previous scan, TH experiences no collision. If some new tags join the population, some of the successful read nodes of the previous scan can now turn into collision nodes and some empty read nodes can turn into successful or collision nodes. If some old tags leave the population, some successful read nodes will become empty read nodes. If any of the new empty read nodes happens to be a sibling of another empty read node, then TH discards these two nodes from the record and stores the location of their parent because the parent is also an empty read node. This strategy works well when the tag population size either remains static or increases. However, when the tag population decreases, the best choice is to reexecute TH for the subsequent scan.

1.5.4 Multiple readers

An application with a large number of RFID tags requires multiple readers with overlapping regions because a single reader can not cover all tags due to the short communication range of tags (usually less than 20 ft). The use of multiple readers introduces several new types of collisions such as reader–reader collisions and reader–tag collisions. Such collisions can be handled by reader scheduling protocols such as those proposed in [5,27,28,31]. TH is compatible with all of these reader scheduling protocols.

1.6 Performance comparison

We implemented two versions of TH. (1) TH_Q, in which γ_{op} is obtained using $E[Q]$ and the query string is matched with MSBs of tag IDs, and (2) TH_T, in which γ_{op} is obtained using $E[T]$ and the query string is matched with LSBs of tag IDs to virtually convert the population distribution into a near-uniform distribution. We also implemented all the 8 prior tag identification protocols in MATLAB®, namely the 3 nondeterministic protocols (Aloha [30], BS [4], and ABS [16]), the 3 deterministic protocols (TW [14], ATW [26], and STT [21]), and the 2 hybrid protocols (MAS [17] and ASAP [22]). As ATW starts DFTs from the level of log z which may not be a whole number, we present results for ATW by both ceiling and flooring the values of log z and representing them with ATW-c and ATW-f, respectively. In terms of implementation complexity, TH and all the 8 prior protocols are implemented in the similar number of lines of code. We performed extensive testing, both manually and automatically, to ensure the correctness of each protocol implementation.

We performed the side-by-side comparison with TH, although this comparison is not completely fair for TH for two reasons. First, 3 of these 8 protocols (i.e., BS, ABS, and ASAP) require modifications to tags and thus do not work with standard C1G2 tags, whereas TH is fully compliant with C1G2. Second, for the framed-slotted Aloha, to its best advantage, we choose the frame size to be the ideal size, which is equal to the tag population size, disregarding the practical limitations on the frame sizes. We choose tag ID length to be the C1G2 standard 64 bits. We performed the comparison for both the uniform case (where the tag population is uniformly distributed in the ID space) and the nonuniform case (where the tag population is not uniformly distributed in the ID space). For the uniform case, we range tag population sizes from 100 to 100,000 to evaluate the scalability of these protocols. For the nonuniform case, we distribute tag populations in blocks where each block is a continuous sequence of tag IDs. We range block sizes from 5 to 1,000. Our motivation for simulating nonuniform distribution in blocks is that in some applications, such as supply chains, tag IDs often come in such blocks when they are manufactured. For each tag population size, we run each protocol 100 times and report the mean. We compare TH with prior protocols from both reader and tag perspectives.

1.6.1 Reader side comparison

For the reader side, we compared TH with the 8 prior protocols based on the following two metrics: (1) normalized reader queries and (2) identification speed. Normalized reader queries is the ratio of the number of queries that the reader transmits to identify a tag population divided by the number of tags in the population. Similarly, identification speed is the total time that the reader takes to identify a tag population divided by the number of tags in that population.

In general, more queries implies more identification time. However, identification time is not strictly in proportion to the number of queries because different queries may take different amounts of time.

For each metric, in Table 1.3, we show the value of TH divided by that for the best prior C1G2 compliant protocol for this metric in the corresponding category of nondeterministic, deterministic, or hybrid. Note that the only prior C1G2 compliant nondeterministic tag identification protocol is the framed-slotted Aloha and the only prior C1G2 compliant hybrid tag identification protocol is MAS. There are 3 prior C1G2 compliant deterministic tag identification protocols: TW, ATW, and STT. We report min, max, and mean for these ratios for tag populations ranging from 100 to 100,000.

For the two metrics defined above, the absolute performance of TH and all prior 8 tag identification protocols is shown in Figures 1.11(a)–1.12(b), for both uniform and nonuniform distributions. Note that for nonuniform distributions, we fix the tag population size to be 5,000 and range the block size from 2 to 1,000.

1.6.1.1   Normalized reader queries

TH_Q reduces the normalized reader queries of the best prior C1G2 compliant nondeterministic, deterministic, and hybrid tag identification protocols by an average of 82%, 50%, and 61%, respectively, for uniformly distributed tag populations. TH_T reduces the normalized reader queries of the best prior C1G2 compliant nondeterministic, deterministic, and hybrid tag identification protocols by an average of 82%, 40%, and 71%, respectively, for nonuniformly distributed tag populations. Figure 1.11(a) and 1.11(b) shows the normalized reader queries of all protocols for uniformly and nonuniformly distributed populations, respectively. Based on these two figures, we make the following four observations from the perspective of normalized reader queries for both uniform and nonuniform distributions. First, normalized queries of TH_T are slightly greater than those of TH_Q for uniformly distributed tag populations. This is because, to minimize identification time, TH_T starts DFTs at levels closer to the leaf nodes compared to TH_Q, which results in more empty reads and less collisions. The increase in number of empty reads is slightly greater than the decrease in number of collisions. Matching the query string with LSBs in TH_T does not bring much advantage because the population is already uniformly distributed. Second, for nonuniformly distributed tag populations, normalized queries of TH_T are, on average, 18% fewer than those of TH_Q. This significant improvement is a result of the virtual conversion of nonuniformly distributed populations into uniformly distributed populations as proposed in Section 1.5.1. Third, among all the 8 prior protocols, the traditional ATW protocol turns out to be the best. Fourth, the framed-slotted Aloha in the C1G2 standard performs the worst even when we disregard the practical limitations on the frame sizes. Although BS is the best among the 3 prior nondeterministic tag

Table 1.3 *Comparison with prior C1G2 compliant protocols (TH/prior art)*

		Prior nondeterministic protocol (= Aloha)			Best prior	Prior deterministic protocols			Prior hybrid protocol (= MAS)		
		Max	Min	Mean		Max	Min	Mean	Max	Min	Mean
Uniform / Prior TH_0/Prior	#queries/tag	0.24	0.10	0.18	ATW-f	0.51	0.50	0.50	0.39	0.38	0.39
	query time/tag	0.84	0.71	0.76	ATW-c	0.92	0.89	0.90	0.81	0.78	0.79
	#responses/tag	0.85	0.59	0.69	ATW-c	0.85	0.67	0.70	0.64	0.24	0.38
	response fairness	1.15	1.10	1.13	TW	1.12	1.07	1.11	1.12	1.07	1.10
Nonuni / TH_T/Prior	#queries/tag	0.26	0.10	0.18	ATW-f	0.75	0.33	0.60	0.40	0.18	0.29
	query time/tag	0.40	0.12	0.24	ATW-f	0.60	0.19	0.41	0.21	0.09	0.15
	#responses/tag	0.63	0.11	0.32	ATW-c	0.87	0.11	0.33	0.46	0.08	0.22
	response fairness	1.38	1.25	1.35	ATW-c	1.03	1.00	1.02	1.05	0.95	1.02

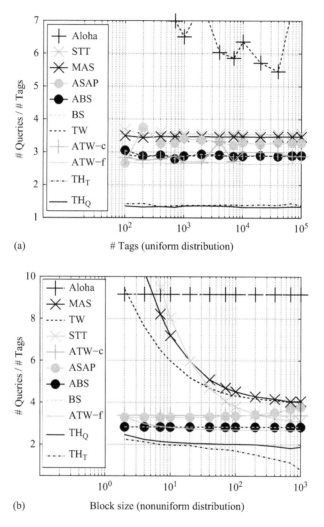

Figure 1.11   Normalized queries of TH and existing protocols: (a) uniform distribution and (b) nonuniform distribution

identification protocols, it is not compliant with C1G2. Similarly, although ASAP is the best among the 2 prior hybrid tag identification protocols, it is not compliant with C1G2.

1.6.1.2   Identification speed

TH_Q improves the identification speed of the best prior C1G2 compliant nondeterministic, deterministic, and hybrid tag identification protocols by an average of 24%, 10%, and 21%, respectively, for uniformly distributed tag populations. TH_T improves

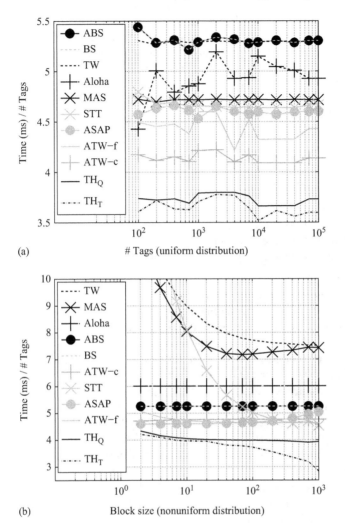

*Figure 1.12   Identification speed of TH and existing protocols: (a) uniform
distribution and (b) nonuniform distribution*

*the identification speed of the best prior C1G2 compliant nondeterministic, deter-
ministic, and hybrid tag identification protocols by an average of 76%, 59%, and
85%, respectively, for nonuniformly distributed tag populations.* Figure 1.12(a) and
1.12(b) shows the identification speed of all protocols for uniformly and nonuniformly
distributed tag populations, respectively. Based on these two figures, we make the
following four observations from the perspective of identification speed. First, the
normalized identification times of TH_T are slightly smaller than those of TH_Q for
uniformly distributed tag populations. This improvement is the result of using $E[T]$
to calculate γ_{op} instead of using $E[Q]$. Second, the normalized identification times of

TH_T are, on average, 36% smaller than those of TH_Q for non-uniformly distributed tag populations. This significant improvement is a result of the virtual conversion of non-uniformly distributed populations into uniformly distributed populations as proposed in Section 1.5.1. Third, among all 8 prior protocols, the traditional ATW protocol turns out to be the best for both uniform and nonuniform distributions. Fourth, although framed-slotted Aloha is the worst in terms of normalized reader queries, its identification speed is not the worst. This is because in our experiments we allow it to use unrealistically large frame sizes, which leads to many empty slots and empty read is much faster than successful read and collision.

1.6.2   Tag side comparison

On the tag side, we compare TH with the 8 prior protocols based on the following four metrics: (1) normalized tag responses, (2) response fairness, (3) normalized collisions, and (4) normalized empty reads. Normalized tag responses is the ratio of sum of responses of all tags during the identification process to the number of tags in the population. Response fairness is the Jain's fairness index given by $\frac{(\sum_{i=1}^{z} x_i)^2}{z \cdot \sum_{i=1}^{z} x_i^2}$ where x_i is the total number of responses by tag i [13]. Normalized collisions is the ratio of total number of collisions during the identification process to the number of tags in the population. Normalized empty reads is the ratio of total number of empty reads during the identification process to the number of tags in the population.

The first two metrics are important for active tags because active tags are powered by batteries. Lesser number of normalized tag responses mean lesser power consumption for active tags. Response fairness measures the variance in the number of responses per tag. Less fairness results in the depletion of the batteries of some tags more quickly compared to others. In large-scale tag deployments, it is often nontrivial to identify tags with depleted batteries and replace them. Using an absolutely fair tag identification protocol, the batteries of all tags deplete at the same time and therefore all can be replaced at the same time. We use the Jain's fairness metric defined in [13]. For z tags, the fairness value is in the range $[\frac{1}{z}, 1]$. The higher this fairness value is, the more fair the protocol is. The second two metrics are important for understanding these identification protocols.

For normalized tag responses and response fairness, in Table 1.3, we show the value of TH divided by that for the best prior C1G2 compliant protocol in the corresponding category of nondeterministic, deterministic, or hybrid. The absolute performance of TH and all prior 8 tag identification protocols is shown in Figures 1.13(a)–1.14(b), for both uniform and nonuniform distributions.

1.6.2.1   Normalized tag responses

TH_Q reduces the normalized tag responses of the best prior C1G2 compliant nondeterministic, deterministic, and hybrid tag identification protocols by an average of 31%, 30%, and 62%, respectively, for uniformly distributed tag populations. TH_T reduces the normalized tag responses of the best prior C1G2 compliant nondeterministic, deterministic, and hybrid tag identification protocols by an average of 68%, 67%, and 78%, respectively, for nonuniformly distributed tag populations. Figure 1.13(a)

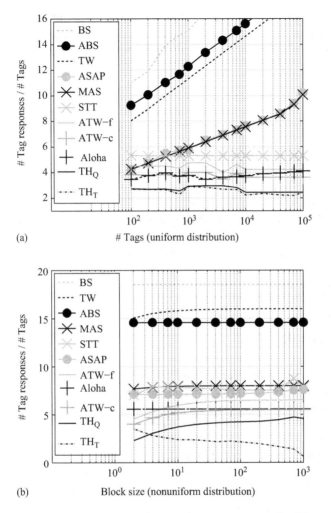

(a)   # Tags (uniform distribution)

(b)   Block size (nonuniform distribution)

Figure 1.13   *Normalized responses of TH and existing protocols: (a) uniform distribution and (b) nonuniform distribution*

and 1.13(b) shows the normalized tag responses of all protocols for uniformly and nonuniformly distributed tag populations, respectively. We make following four observations from these two figures. First, the normalized tag responses of TH_T are, on average, 57% lesser than those of TH_Q for nonuniformly distributed tag populations. Second, the normalized tag responses of BS, ABS, TW, MAS, and ASAP increase with increasing tag population size. Third, for nonuniformly distributed tag populations, the normalized tag responses of nondeterministic protocols is not affected by the block size because their performance is independent of tag ID distribution. In contrast, the normalized tag responses of deterministic protocols slightly increase with increasing block size. Fourth, among all 8 prior protocols, Aloha has the smallest number of normalized tag responses. This is because of the unlimitedly large frame sizes that we used for Aloha. With large frame sizes, tags experience lesser collisions and thus reply fewer times.

1.6.2.2 Tag response fairness

TH_Q improves the tag response fairness of the best prior C1G2 compliant nondeterministic, deterministic, and hybrid tag identification protocols by an average of 13%, 11%, and 10%, respectively, for uniformly distributed tag populations. TH_T improves the tag response fairness of the best prior C1G2 compliant nondeterministic, deterministic, and hybrid tag identification protocols by an average of 35%, 2%, and 2%, respectively, for non-uniformly distributed tag populations. Figure 1.14(a) and 1.14(b) shows the tag response fairness of all protocols for uniformly and nonuniformly distributed tag populations, respectively. We observe that among all 8 prior protocols, ASAP and ATW are the best for uniformly and nonuniformly distributed populations, respectively. We observe that TH_T achieves slightly better fairness than TH_Q.

Figure 1.17(a) and 1.17(b) shows the distribution of the number of tag responses for each protocol for uniformly and nonuniformly distributed tag populations, respectively. For any protocol, the wider the horizontal span of its distribution is, the larger the range of the number of responses per tag it has. We observe that TH has the smallest range among all protocols for the number of responses per tag.

1.6.2.3 Normalized collisions

TH_Q and TH_T both incur smaller number of collisions than all 8 prior protocols for uniformly and non-uniformly distributed tag populations. Figure 1.15(a) and 1.15(b) shows the normalized collisions for all protocols for uniformly and nonuniformly distributed tag populations, respectively. From these figures, we make following three observations. First, TH_T incurs fewer collisions compared to TH_Q, which is one of the reasons behind the faster identification speed of TH_T. Second, Aloha incurs the smallest number of normalized collisions among all 8 prior protocols because of the unlimitedly large frame sizes that we used for it. Third, TW mostly incurs the largest number of normalized collisions for both types of populations.

1.6.2.4 Normalized empty reads

For uniformly distributed tag populations, TH_Q incurs a smaller number of empty reads than all 8 prior protocols. For nonuniformly distributed tag populations, TH_T, incurs a smaller number of empty reads than all 8 prior protocols. Figure 1.16(a) and 1.16(b) shows the normalized empty reads of all protocols for uniformly and nonuniformly distributed tag populations, respectively. From these figures, we observe that although the two prior C1G2 compliant protocols, TW and MAS, have fewer empty reads compared to TH_Q for large block sizes, they have much larger number of collisions compared to TH_Q, which makes their overall identification time much larger than TH_Q. Note that the slightly larger number of empty reads for TH_Q for large block sizes is immaterial because the time for an empty read is 5 times lesser than that for a collision and 10 times lesser than that for a successful read. Therefore, reducing the number of collisions is more important than reducing the number of empty reads. We also observe that TH_T has greater number of empty reads compared to TH_Q, which

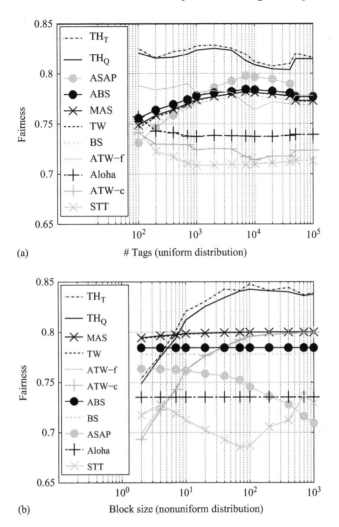

Figure 1.14 Response fairness of TH and existing protocols: (a) uniform distribution and (b) nonuniform distribution

is the cost of decreasing the collisions. As collisions are 5 times slower compared to empty reads, this slight increase in number of empty reads is not of much significance.

Note that the collisions and empty reads shown in Figures 1.15(a) and 1.16(b), respectively, are consistent with the reader queries shown in Figure 1.11(a) as well as the identification speed shown in Figure 1.12(a). Similarly, the collisions and empty reads shown in Figures 1.15(b) and 1.16(b), respectively, are consistent with the reader queries shown in Figure 1.11(b) as well as the identification speed shown in Figure 1.12(b). For example, Figure 1.15(a) shows that TW has more collisions than Aloha, but Figure 1.11(a) shows that Aloha has more queries than TW. This is because

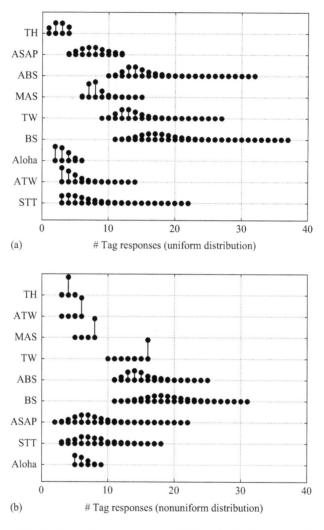

Figure 1.15   Distribution of tag responses of TH and existing protocols: (a) uniform distribution and (b) nonuniform distribution

Aloha has much more empty reads than TW as shown in Figure 1.16(a). Although Aloha has more queries than TW, Figure 1.12(a) also shows that Aloha requires less identification time than TW. This is because an empty read is five times faster than a collision for a reader.

A common observation that we make from the plots of all the metrics of TH for uniformly distributed populations is that these plots have ups and downs and are not monotonic. This is because when the number of tags increases, the starting level from where TH performs the first DFT increases, which has an effect on all these metrics. These ups and downs are also observed in the analytical plot in Figure 1.8.

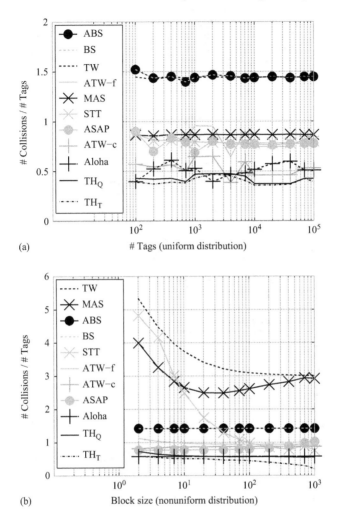

Figure 1.16 Normalized collisions of TH and existing protocols: (a) uniform distribution and (b) nonuniform distribution

1.7 Conclusion

The technical novelty of this chapter lies in that it represents the first effort to formulate the TW process mathematically and propose a method to minimize the expected number of queries and expected identification time. The significance of this chapter in terms of impact lies in that the TW protocol is a fundamental multiple access protocol and has been standardized as an RFID tag identification protocol. Besides static optimality, our TH protocol dynamically chooses a new optimal level after each subtree is traversed. We presented a method to make our protocol work with nonuniformly distributed populations and achieve similar performance that it achieves with

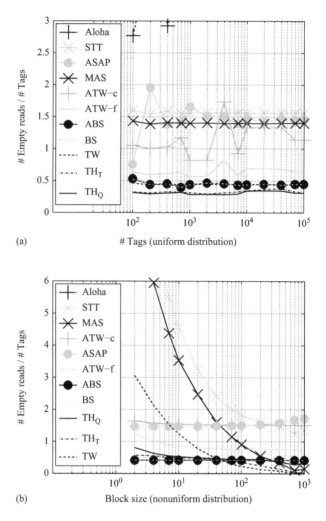

Figure 1.17   Normalized empty reads of TH and existing protocols: (a) uniform distribution and (b) nonuniform distribution

uniformly distributed populations. We also presented methods to make our protocol reliable, to continuously scan tag populations that are dynamically changing, and to work with multiple readers with overlapping regions. Another key contribution of this chapter is that we conducted a comprehensive side-by-side comparison of two variants of our protocol with eight major prior tag identification protocols that we implemented. Our experimental results show that our protocol significantly outperforms all prior tag identification protocols, even those that are not C1G2 compliant, for metrics such as the number of reader queries per tag, the identification speed, and the number of responses per tag.

Chapter 2
RFID identification—fairness

2.1 Introduction

2.1.1 Motivation and problem statement

Radio-frequency-identification (RFID) system is a wireless system that is comprised of readers and tags. The job of the reader is to read the IDs of the tags and possibly some data appended to the IDs. The job of the tag is to transmit its ID and the appended data to the reader when asked for. A reader has a dedicated power source with significant computing power. It reads the IDs of tags and arbitrates the shared wireless medium among them. There are two types of tags: (1) passive, which do not have batteries, are powered up by harvesting the radio frequency energy from readers and have communication ranges often less than 20 ft and (2) active, which have batteries and have communication ranges up to 300 ft.

RFID systems with active tags are widely used in various applications, such as supply chain management [32], object tracking [19], inventory control, electronic toll collection, and access control [18], because active tags have better communication quality and reliability than passive tags. For example, Walmart uses active tags to track expensive clothing merchandize for better inventory control [33], and Honeywell Aerospace uses active tags to track its products from birth through use by airlines and repair by Honeywell [34]. Active tags with embedded sensors have seen increasing adoption in applications such as health care [35], supply chain management [32], monitoring server cabinet temperatures [36], and disaster management [37]. An example of active tags with embedded sensors is the Tempcorder series of RFID tags used for underground railways monitoring, industrial hazard prevention, and data center monitoring [38]. In this chapter, we deal with active RFID tags.

In the routine task of reading the IDs of active tags, i.e., the process of RFID identification, tags transmit their IDs to readers over a shared wireless medium. Therefore, their transmissions often collide causing some tags to use their scarce energy resources to retransmit their IDs. The transmission of IDs to the readers is the most energy-consuming activity of an active tag [39]. RFID tags transmit and receive at a constant power regardless of how close or far they are from the reader. In many scenarios such as railways monitoring, reflection seismology (used by oil exploration companies to determine the subsurface rock structure), and automotive and aircraft assembly lines, the tags have to transmit their IDs every few seconds [40–43]. Due to the frequent transmissions, batteries of the tags used in such applications have a lifetime only of the order of a few weeks. If the RFID identification protocol is unfair in the sense that some tags transmit more times than others, then the batteries of those tags deplete faster than others. When the batteries of tags do not deplete at the same time, every few weeks, network operators have to identify and locate tags that have depleted batteries so that they can replace their batteries, which is typically a manual and laborious process. In many settings, it can be hard to find tags with depleted batteries because they cannot be electronically located anymore. In contrast, if the number of transmissions per tag during identification is roughly equal for all tags, then tag batteries will deplete at roughly the same time and thus the network operators can replace the batteries of all tags at the same time.

This chapter addresses the fundamental problem of achieving fairness in RFID identification while minimizing identification time. Specifically, *given a tag population of unknown size and a required fairness α, design a tag identification protocol that minimizes the identification time under the constraint that the fairness in terms of number of transmissions per tag is no less than α.* We use the well-known Jain's fairness index to quantify fairness [13]. The identification protocol should be compliant with the prevalent EPCGlobal Class 1 Generation 2 (C1G2) RFID standard [53].

2.1.2   Limitations of prior art

To the best of our knowledge, no prior work has been targeted on developing a fair RFID identification protocol that can achieve any required fairness. Existing RFID identification protocols mostly focus on minimizing identification time and are unfair because in each round of identification, the *same* set of tags has to transmit more times compared to the other tags. In [44], Shahzad and Liu evaluated the fairness of their TH protocol along with other RFID identification protocols, and showed that TH achieves the highest fairness among all existing protocols. However, TH is not designed to achieve any required fairness; its design objective is to just minimize the identification time without fairness constraints. In contrast, our objective is to minimize the identification time with fairness constraints. The maximum value of Jain's fairness Index that TH achieves is only 0.84. To understand what fairness of 0.84 means, consider a population of active tags, where the battery of each tag lasts for a hundred thousand transmissions, and a reader performs one thousand identification rounds each day. For a protocol with fairness of 0.84, the batteries of 20% tags will last for 33 days, of another 30% tags will last for 50 days, and of the remaining 50% will last for about 100 days. In comparison, with fairness of 0.99, the batteries of 99% tags will last for 100 days, and only 1% will deplete in 50 days. Even though the batteries of 1% tags will deplete in 50 days, they won't usually need to be found and replaced because in large-scale deployments, e.g., in railways monitoring, loss of up to 10% tags is tolerable because the remaining tags will still provide data that is enough to ensure smooth operation. This tolerability of 10% is application specific, but usually in most applications, the loss of up to 5% tags is tolerable.

2.1.3 Proposed approach

In this chapter, we propose FRIP, the first $\underline{F}$air $\underline{R}$FID $\underline{I}$dentification $\underline{P}$rotocol. The communication protocol used by FRIP is the C1G2 standardized frame-slotted Aloha protocol, in which a reader first broadcasts a value f to the tags in its vicinity, where f represents the number of time slots present in a forthcoming frame and lies in the range $1 \leq f \leq 2^{15}$. Each C1G2 compliant tag has an inventory bit, which is initialized to 0. Each tag whose inventory bit is 0 randomly picks a time slot in the frame and transmits during that slot. If no tag transmits in a slot, it is called an *empty slot*; if exactly one tag transmits, it is called a *singleton slot*; and if two or more tags transmit, it is called a *collision slot*. In a singleton slot, the reader successfully gets the ID of the transmitting tag and issues a command to the tag to change its inventory bit to 1. The reader executes several frames until all slots in the frame are empty slots, which indicates that all tags have been identified. We assume that the communication channel between readers and tags is reliable i.e., tags correctly receive queries from the readers and the readers correctly receive responses from the tags. We further assume that each tag in the population is covered by at least one reader during the identification rounds.

The key idea behind FRIP is to bind the expected number of tags that transmit more than once by finding the optimal frame sizes. Such frame sizes ensure that the required fairness is achieved while minimizing the identification time. *FRIP optimally trades-off the identification time for fairness.*

2.1.3.1 Identification process

FRIP performs identification in three steps. First, it estimates the size of the RFID tag population. Second, based on the estimate, it calculates the optimal frame size and executes an Aloha frame of this size. Third, after executing the frame, it re-estimates the size of remaining unidentified tag population, uses it to recalculate a new optimal frame size, and executes another frame. FRIP repeats this until it has identified all tags.

Population size estimation: To estimate the size of tag population, FRIP uses ART, an estimation scheme proposed by Shahzad and Liu in [45]. We chose ART because of its proven reliability, speed, C1G2 compliance, and ability to estimate sizes of arbitrarily large tag populations. ART uses the frame-slotted Aloha protocol and thus, integrates seamlessly with FRIP. To obtain an estimate, ART executes several Aloha frames and obtains a binary sequence of 0s and 1s from each frame by representing empty slots in the frame with 0s and singleton and collision slots with 1s. It then calculates the average run size of 1s in the binary sequences, which is a monotonically increasing function of the size of tag population, and uses it to estimate the size of tag population.

Calculating optimal frame size: To calculate the optimal frame size, we first express Jain's fairness index as a function of the expected number of singleton slots in each frame. We then express the expected number of singleton slots in any given frame as a function of ratio of unidentified tag population size and frame size. Using these two expressions, FRIP calculates an upper bound for the ratio of unidentified tag population size and frame size that should not be exceeded in order to achieve the required fairness. It then calculates the optimal value of this ratio that minimizes the expected identification time under the constraint that the ratio does not exceed the upper bound. As the estimate of unidentified tag population size is known at the start of each frame, FRIP uses the optimal value of the ratio to calculate the optimal size for each frame. FRIP calculates the optimal value of the ratio *only once* at the start of the identification process and uses this value to calculate all frame sizes throughout the process.

Reestimation: After each frame, FRIP estimates the size of remaining unidentified tag population by subtracting the number of successfully identified tags at the end of the frame from the number of unidentified tags at the start of that frame. Note that FRIP does not execute ART again to reestimate the size of tag population. It executes ART only once at the start of the identification process. It recalculates the optimal size for the next frame by dividing the optimal value of the ratio by the number of remaining unidentified tags and executes the next frame. It continues this process until all slots in a frame turn out as empty slots.

2.1.3.2 Large frame size implementation

For large populations, the value of the optimal frame size may exceed the upper limit of 2^{15} slots per frame specified in the C1G2 standard. If this happens, FRIP uses SELECT command, standardized in the C1G2 standard, to specify a subset of tag population that should participate in the identification process in any given frame. Thus, FRIP *virtually* divides the entire population into smaller groups of roughly equal sizes and then identifies the tags in each group independently. This allows FRIP to use frame sizes $\leq 2^{15}$.

2.1.3.3 Multiple readers

An application with a large number of RFID tags requires multiple readers with overlapping regions because a single reader may not be able to cover all tags. The use of multiple readers introduces several new types of collisions such as reader-reader collisions and reader-tag collisions. Such collisions can be handled by reader scheduling protocols such as those proposed in [27,31]. FRIP is compatible with all of these reader scheduling protocols.

2.1.4 Key novelty and contributions

The key novelty of this chapter lies in proposing FRIP, an RFID identification protocol that statistically guarantees to achieve any required fairness. The key contributions of this chapter are in the mathematical development behind FRIP in making it fair as well as the extensive evaluation and comparison of FRIP with existing protocols. For mathematical development, we formulated the fair identification problem as a constraint optimization problem and solved it optimally. For evaluation and comparison, we implemented FRIP and nine RFID identification protocols, which represent the state-of-the-art, in MATLAB®. These nine protocols are Aloha [30], BS [4], ABS [16], TH [44], TW [14], ATW [26], STT [21], MAS [17], and ASAP [22].

We performed side-by-side comparisons of FRIP with these protocols using tag populations ranging in sizes from a hundred tags to a million tags and following different distributions. Our results show that FRIP always achieves the required fairness, whereas the fairness value of the best existing protocol, TH, is only 0.84. FRIP reduces the average number of transmissions per tag by at least 2.62 times compared to the best existing protocol. It also reduces the average energy consumption per tag by at least 11% compared to the best existing protocol. In terms of identification time, FRIP is faster than all existing protocols except TH when the values of required fairness in FRIP are set equal to the fairness values achieved by the corresponding existing protocols. Although TH is faster than FRIP by 13.9%, it cannot achieve arbitrarily high fairness.

2.2 Related work

Ferrero *et al.* proposed fair neighbor friendly reader anticollision protocol (FNFRA) that focuses on increasing fairness in throughput of RFID readers in multiple readers' environment [46]. FNFRA increases the fairness in throughput by appropriately adjusting the duration of time each reader gets to identify tags in its coverage region such that the number of tags each reader identifies in its allocated time duration is equal. Note that the objectives of FNFRA and FRIP are fundamentally different: FNFRA focuses on making reader scheduling process fair, while FRIP focuses on making the tag identification process fair. To the best of our knowledge, no work has been done towards proposing a protocol to make the tag identification process fair. Existing identification protocols mostly focus on minimizing the identification time and can be divided into three types: (1) deterministic, (2) nondeterministic, and (3) hybrid.

2.2.1 Deterministic identification protocols

Existing such protocols are based on TW protocol [14]. In TW, a reader first queries 0 and all tags whose IDs start with 0 transmit. If result of the query is a successful read (i.e., exactly one tag transmits) or an empty read (i.e., no tag transmits), the reader queries 1 and all tags whose IDs start with 1 transmit. If the result of the query is a collision, the reader generates two new query strings by appending a 0 and a 1 to the previous query string and queries the tags with these new query strings. All the tags whose IDs start with the new query string transmit. This process continues until all the tags are identified. TW-based protocols incur large number of collisions resulting in lack of fairness in each round of identification. Note that if the tag population does not change significantly, the tags that experienced comparatively more collisions in previous round of identification experience more collisions in the next round of identification as well. Therefore, the batteries of the tags that experience more collisions in each round deplete at a faster rate compared to the batteries of the tags that experience less collisions. Although several protocols, such as STT [21],

ATW [26], and TH [44], have been proposed to reduce the number of collisions, they focus on reducing the identification time and do not address the fairness issue. Token-MAC is a deterministic protocol, not based on TW, in which a reader provides tokens, i.e., permission to transmit, to tags such that tags with lower historical success rate receive more tokens [47].

2.2.2   *Nondeterministic identification protocols*

Existing such protocols are either based on frame-slotted Aloha [30] or Binary Splitting (BS) [4]. Frame-slotted Aloha was described in Section 2.1.3. In BS, the reader asks the tags to transmit their IDs. If more than one tag transmit, BS randomly divides and subdivides the population into smaller groups until each group has only one or no tag. This process of subdivision incurs a lot of collisions resulting in lack of fairness in tag transmissions. ABS is a BS-based protocol that is designed for continuous identification of tags [16]. CP is another nondeterministic protocol in which tags cooperate and relay transmissions of neighboring tags to the reader [48]. CP is not based on Aloha or BS.

2.2.3   *Hybrid identification protocols*

Hybrid protocols combine features from both nondeterministic and deterministic protocols. There are two major such protocols: multislotted scheme with assigned slots (MAS) [17] and adaptively splitting-based arbitration protocol (ASAP) [22]. MAS is a TW-based protocol in which each tag that matches the reader's query picks up one of the f time slots to transmit. ASAP divides and subdivides the tag population into subsets until the size of each subset is below a certain threshold and then applies Aloha on each subset. Both these protocols also aim to minimize the identification time and do not address the fairness issue.

2.3   Optimal frame size

FRIP calculates optimal frame size at the start of each frame to ensure that the required amount of fairness in number of transmissions per tag is achieved. It uses the estimate of the unidentified tag population size at the start of each frame to calculate the optimal frame size. To obtain expressions for calculating the optimal frame size, we first express Jain's fairness index as a function of the expected number of singleton slots in each frame. Second, we express the time to identify all tags in the population in terms of number of singleton slots, collision slots, and empty slots in each frame. Third, we express the expected number of singleton, collision, and empty slots in any given frame as a function of the ratio of unidentified tag population size and frame size. Fourth, we calculate the expected number of frames that FRIP has to execute to identify all tags in the population. Last, we substitute the expressions of expected values obtained from third and fourth steps into the expressions obtained from first and second steps, and formulate the problem as a constraint optimization problem. The objective of this optimization problem is to minimize the expected identification time

Table 2.1 Symbols used in the chapter

Symbol	Description
t	Estimated # of tags in the population
t_i	# of unidentified tags at start of ith frame
f_i	Size of ith frame
f_{max}	Max. value of 2^{15} for f_i allowed by the C1G2
$\mathscr{J}$	Jain's fairness index
α	Required fairness
x_l	# of times tag with label l transmits
s_i, c_i, e_i	Singleton, collision, and empty slots in ith frame
S_i, C_i, E_i	Random var. for # of singleton, collision, and empty slots in ith frame
X_j	Indicator random var. for jth slot to be singleton
Y_j	Indicator random var. for jth slot to be empty
n	Expected # of frames to identify all tags
T_s, T_c, T_e	Time duration of singleton, coll., and empty slots
$\mathscr{T}$	Total identification time
k	t_i/f_i
k_{min}	Value of k for which $E[\mathscr{T}]$ is minimum
m	Upper bound on k
μ	KKT multiplier
$E[.]$	Expected value operator

under the constraint that the fairness in the number of transmissions per tag is at least equal to the required fairness. More specifically, using the substituted expression from the first step, we calculate an upper bound on the ratio of unidentified tag population size and frame size. Then using the substituted expression from the second step, we calculate the optimal value of this ratio that minimizes the expected identification time under the constraint that the ratio does not exceed the upper bound. As the estimate of unidentified tag population size is known at the start of each frame, FRIP divides it by the optimal value of this ratio to calculate the optimal size for each frame. FRIP is computationally inexpensive because it calculates the optimal value of the ratio only once at the start of the identification process and uses it for all frames throughout the identification process. Next, we explain these five steps followed by a strategy to execute frame sizes that are larger than 2^{15}. Table 2.1 lists the symbols used in this chapter.

2.3.1 Jain's fairness index

Jain *et al.* introduced an index, called Jain's fairness index, to rate the fairness in allocation of a resource to a set of users/devices [13]. This index has become the most commonly used metric to measure fairness in communication systems. Let t represent the total number of devices sharing a resource and x_l be the portion of the resource allocated to lth device. The expression for Jain's fairness index is

$$\mathscr{J} = \frac{\left(\sum_{l=1}^{t} x_l\right)^2}{t \times \sum_{l=1}^{t} x_l^2} \tag{2.1}$$

The value of this fairness index ranges from $1/t$ (worst case when one user gets all the resource and remaining $t - 1$ users get nothing) to 1 (best case when all users get equal share of the resource).

We adapt Jain's fairness index in (2.1) to measure fairness in the number of transmissions per tag by expressing it as a function of the expected number of singleton slots in each frame. Let t represent the total number of tags in the population. We arbitrarily assign a unique label l to each tag, where $1 \leq l \leq t$. Let x_l represent the number of times tag with label l transmits its ID during the identification process. Ideally the value of x_l should be same for all values of l, which will result in maximum fairness of 1. Recall that a reader executes several Aloha frames to identify all tags in the population. Let s_i represent the number of singleton slots in ith frame, i.e., s_i tags are successfully identified in ith frame. As per the Aloha protocol, a tag transmits its ID once in each frame until it is successfully identified. This means that a tag that is identified in the ith frame, transmitted its ID i times. Let n represent the expected number of frames that the reader will execute to identify all the tags in the population. The summation term in the numerator of (2.1), which gives the sum of number of transmissions by each of the t tags during the identification process, becomes $\sum_{l=1}^{t} x_l = s_1 \times 1 + s_2 \times 2 + s_3 \times 3 + \cdots + s_n \times n$. Similarly, the summation term in the denominator of (2.1), which gives the sum of square of number of transmissions by each tag, becomes $\sum_{l=1}^{t} x_l^2 = s_1 \times 1^2 + s_2 \times 2^2 + s_3 \times 3^2 + \cdots + s_n \times n^2$. Thus, Jain's fairness index can be represented as

$$\mathscr{J} = \frac{\left(\sum_{i=1}^{n} s_i \times i\right)^2}{t \left(\sum_{i=1}^{n} s_i \times i^2\right)} \tag{2.2}$$

2.3.2   Total identification time

We express the time to identify all tags in a population of size t in terms of number of singleton slots, collision slots, and empty slots in each frame. The total identification time is given by the sum of the time taken by singleton slots, collision slots, and empty slots in all n frames. Let s_i, c_i, and e_i represent the number of singleton, collision, and empty slots, respectively, in ith frame. Let T_s, T_c, and T_e be the absolute time durations of singleton, collision, and empty slots, respectively. The total identification time $\mathscr{T}$ is given by

$$\mathscr{T} = \sum_{i=1}^{n} (s_i \times T_s + c_i \times T_c + e_i \times T_e) \tag{2.3}$$

where $s_i + c_i + e_i = f_i$, and f_i is the size of ith frame.

2.3.3   Expected values of slots

We express the expected values of singleton, collision, and empty slots in ith frame as a function of ratio t_i/f_i, where t_i is the size of unidentified tag population at the start of the ith frame and f_i is the size of the ith frame. Before formally deriving expressions for these expected values, we state an assumption that we use to make the formal development tractable.

We assume that instead of picking a single slot to transmit at the start of ith frame of size f_i, a tag independently decides to transmit in each slot of the frame with probability $1/f_i$ regardless of its decision about previous or forthcoming slots. Vogt first used this assumption for the analysis of frame-slotted Aloha protocol for RFID and justified its use by recognizing that this problem belongs to a class of problems called *occupancy problem*, which deals with the allocation of balls to urns [49]. Ever since, the use of this assumption has become a norm in the formal analysis of all Aloha based RFID protocols [30,45,49,50].

The implication of this assumption is that a tag can end up choosing more than one slots in the same frame or even not choosing any at all, which is not in accordance with the C1G2 standard that requires a tag to pick exactly one slot in a frame. However, this assumption does not create any problems because the expected number of slots that a tag chooses in a frame is still one. The analysis with this assumption is, therefore, asymptotically the same as that without this assumption. Bordenave *et al.* further explained in detail why this independence assumption in analyzing Aloha-based protocols provides results just as accurate as if all the analysis was done without this assumption [51]. Note that this independence assumption is made only to make the formal development tractable. All simulations in the chapter are based on the C1G2 standard where a tag chooses exactly one slot at the start of each frame.

Lemma 2.1 gives the expressions for expected value of number of singleton, collision, and empty slots.

Lemma 2.1. *Let S_i, C_i, and E_i be the random variables representing number of singleton, collision, and empty slots, respectively, in the ith frame. Let the size of the ith frame be f_i. Let t_i represent the number of unidentified tags in the population at the start of the ith frame. Let k be the ratio of t_i to f_i for all $1 \leq i \leq n$, i.e., $k = t_i/f_i$. Then*

$$E[S_i] = kf_ie^{-k} \tag{2.4}$$

$$E[C_i] = f_i - kf_ie^{-k} - f_ie^{-k} \tag{2.5}$$

$$E[E_i] = f_ie^{-k} \tag{2.6}$$

Proof. Let j be any arbitrary slot in the frame, where $1 \leq j \leq f_i$. Let X_j be an indicator random variable for the event when jth slot is a singleton slot, i.e., $X_j = 1$ if exactly one tag transmits in the jth slot, otherwise $X_j = 0$. The value of S_i in terms of X_j is therefore given by $S_i = \sum_{j=1}^{f_i} X_j$. The number of tags that transmit in any given slot of a frame follow a binomial distribution. Therefore,

$$E[X_j] = P\{X_j = 1\} = \binom{t_i}{1}\left(\frac{1}{f_i}\right)\left(1 - \frac{1}{f_i}\right)^{t_i-1} \approx ke^{-k} \tag{2.7}$$

As $\{X_1, X_2, \ldots, X_{f_i}\}$ forms a set of identically distributed random variables, the expected value of S_i is given by $E[S_i] = E\left[\sum_{j=1}^{f_i} X_j\right] = f_i \times E[X_j] = kf_ie^{-k}$.

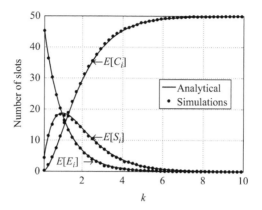

Figure 2.1 s_i, c_i, and e_i vs. k

Similarly, let Y_j be an indicator random variable for the event when jth slot is an empty slot, which implies that $E_i = \sum_{j=1}^{f_i} Y_j$. Therefore,

$$E[Y_j] = P\{Y_j = 1\} = \binom{t_i}{0}\left(\frac{1}{f_i}\right)^0 \left(1 - \frac{1}{f_i}\right)^{t_i} \approx e^{-k} \tag{2.8}$$

Thus, the expected value of E_i becomes $E[E_i] = E\left[\sum_{j=1}^{f_i} Y_j\right] = f_i \times E[Y_j] = f_i e^{-k}$. As $C_i = f_i - S_i - E_i$, $E[C_i] = f_i - E[S_i] - E[E_i] = f_i - kf_i e^{-k} - f_i e^{-k}$.   □

We use the expressions for $E[S_i]$, $E[C_i]$, and $E[E_i]$ as values of s_i, c_i, and e_i, respectively, i.e., $s_i = E[S_i]$, $c_i = E[C_i]$, and $e_i = E[E_i]$. Figure 2.1 shows the theoretically calculated expected values of number of singleton, collision, and empty slots from (2.4) to (2.6), and experimentally observed values of s_i, c_i, and e_i from simulations. The continuous lines in the figure are the theoretical plots and the dots are the experimentally observed values from simulations. We make two conclusions from Figure 2.1. First, our independence assumption does not cause the theoretical analysis to deviate from practically observed values. Second, we can use $s_i = E[S_i]$, $c_i = E[C_i]$, and $e_i = E[E_i]$.

2.3.4   Expected number of Aloha frames

Next, we calculate the expected number of frames n that FRIP executes to identify all tags in the RFID tag population. Lemma 2.2 gives the expression for n.

Lemma 2.2. *Let k be the ratio of number of unidentified tags in the population at the start of any given frame to the size of that frame. Let size of the population be t. The expected number of frames, n, that FRIP executes to identify these t tags is given by:*

$$n = -\frac{\ln\{t\}}{\ln\{1 - e^{-k}\}} \tag{2.9}$$

Proof. Recall that t_i represents the number of unidentified tags at the start of ith frame. The value of t_i can be calculated iteratively using the equation $t_i = t_{i-1} - s_{i-1}$, where $t_1 = t$. As $s_{i-1} = E[S_{i-1}] = kf_{i-1}e^{-k} = \frac{t_{i-1}}{f_{i-1}}f_{i-1}e^{-k} = t_{i-1}e^{-k}$. Therefore, $t_i = t_{i-1} - t_{i-1}e^{-k} = t_{i-1}(1 - e^{-k})$. It is straightforward to see that the closed form expression for t_i is

$$t_i = t(1 - e^{-k})^{i-1} \tag{2.10}$$

The value of i for which $t_i = 1$ is the expected value of the total number of frames n that FRIP executes because $t_i = 1$ means that there is only one unidentified tag in the population at the start of the ith frame and it will be identified without any collisions in the ith frame. Therefore, the value of n can be obtained by equating t_n with 1, i.e., $t_n = t(1 - e^{-k})^{n-1} = 1$, solving this expression for n, and subtracting 1 from it results in (2.9). We subtract 1 because last frame is essentially a single slot with one unidentified tag. We don't count that frame towards total number of frames, instead we will add time equivalent to one singleton slot in the expected identification time to cater for the last single slot frame.   □

2.3.5   Calculating optimal frame size

We substitute the expressions of expected values of s_i, c_i, e_i, and n from (2.4), (2.5), (2.6), and (2.9), respectively, into the expressions for Jain's fairness index and total identification time in (2.2) and (2.3), respectively, to formulate the constraint optimization problem. We first derive the constraint that k must not violate to ensure that observed value of $\mathscr{J}$ is greater than or equal to α. This constraint is in the form of an upper bound on k that k should not exceed. After this, we derive a closed form expression for expected identification time. Using this closed-form expression and the upper bound on k, we formulate the fair identification problem as a constraint optimization problem and solve it to calculate the optimal value of k that minimizes the expected identification time under the constraint that k does not exceed the upper bound. FRIP uses this optimal value of k and divides the estimate of the size of unidentified tag population at the start of each frame by it to calculate the optimal size for each frame. Recall from Section 2.1.3 that FRIP obtains the estimate of size of unidentified tag population at the start of each frame by simply subtracting the number of tags successfully identified in the last frame from the number of unidentified tags at the start of the last frame. Theorem 2.1 calculates the value of the upper bound on k.

Theorem 2.1. *The fairness achieved by FRIP in the number of transmissions per tag will be no less than α if the ratio, k, of number of unidentified tags to frame size is no greater than the upper bound m, where m is given by the numerical solution of the following equation:*

$$\frac{(e^m(t-1)\ln\{A\} + \ln\{t\})^2}{t\left(e^m(2e^m - 1)(t-1)\ln^2\{A\} + 2e^m\ln\{A\}\ln\{t\} - \ln^2\{t\}\right)} = \alpha \tag{2.11}$$

where $A = 1 - e^{-m}$.

Proof. The observed value of the fairness index, $\mathscr{J}$, should satisfy the condition $\mathscr{J} \geq \alpha$. We first express the Jain's fairness index as a function of k and discuss some of its properties. Then we show that if k is less than or equal to m, the required fairness, α, will be achieved i.e., the requirement $\mathscr{J} \geq \alpha$ will be satisfied.

By substituting the value of s_i from (2.4) into (2.2), we get

$$\mathscr{J} = \frac{\left(\sum_{i=1}^{n} ikf_i e^{-k}\right)^2}{t\left(\sum_{i=1}^{n} i^2 kf_i e^{-k}\right)} \tag{2.12}$$

According to (2.10), $t_i = t(1 - e^{-k})^{i-1}$. For simplifying the presentation, we replace $1 - e^{-k}$ by r. Thus, $t_i = tr^{i-1}$. As $k = \frac{t_i}{f_i} \Rightarrow t_i = kf_i$, therefore, $kf_i = tr^{i-1}$. By substituting this value of kf_i into (2.12) and simplifying, we get

$$\mathscr{J} = \frac{e^{-k}\left(\sum_{i=1}^{n} ir^{i-1}\right)^2}{\sum_{i=1}^{n} i^2 r^{i-1}} \tag{2.13}$$

The values of the two summation terms of the equation above are given in the following two equations.

$$\sum_{i=1}^{n} ir^{i-1} = \frac{1}{(r-1)^2}\left\{1 + nr^{n+1} - r^n(n+1)\right\} \tag{2.14}$$

$$\sum_{i=1}^{n} i^2 r^{i-1} = \frac{1}{(r-1)^3}\left\{-1 - r + n^2 r^{n+2} + r^n(n+1)^2\right.$$
$$\left. - r^{n+1}(2n^2 + 2n - 1)\right\} \tag{2.15}$$

Next, we derive these two equations. The expanded form of left hand side of (2.14) is $\sum_{i=1}^{n} ir^{i-1} = 1r^0 + 2r^1 + 3r^2 + \cdots + nr^{n-1}$. Multiplying it with r, and subtracting the resultant from $\sum_{i=1}^{n} ir^{i-1}$, we get

$$\sum_{i=1}^{n} ir^{i-1} - r\sum_{i=1}^{n} ir^{i-1} = \sum_{i=1}^{n} r^{i-1} - nr^n \tag{2.16}$$

Substituting $\sum_{i=1}^{n} r^{i-1}$ by $(1 - r^n)/(1 - r)$ in the equation above and rearranging, we get (2.14). Equation (2.15) can be derived using same steps.

These two equations can be further simplified by putting $r^n = 1/t$. Next we show that r^n is indeed equal to $1/t$. As $r = 1 - e^{-k}$, raising both sides to the power n, using the value of n from (2.9), and taking natural log, we get

$$\ln\{r^n\} = \ln\{(1 - e^{-k})^n\} = n \times \ln\{(1 - e^{-k})\}$$
$$= -\frac{\ln\{t\}}{\ln\{1 - e^{-k}\}} \times \ln\{(1 - e^{-k})\} = \ln\left\{\frac{1}{t}\right\}$$
$$\Rightarrow r^n = \frac{1}{t} \tag{2.17}$$

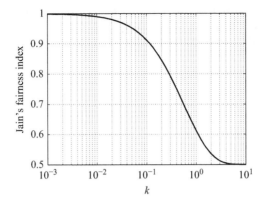

Figure 2.2 $\mathscr{J}$ vs. k

Substituting $r^n = 1/t$ and the value of n from (2.9) into (2.14) and (2.15) and then substituting the resulting expressions of $\sum_{i=1}^{n} i r^{i-1}$ and $\sum_{i=1}^{n} i^2 r^{i-1}$ into (2.13), we get

$$\mathscr{J} = \frac{\left(e^k(t-1)\ln\{r\} + \ln\{t\}\right)^2}{t\left(e^k(2e^k-1)(t-1)\ln^2\{r\} + 2e^k\ln\{r\}\ln\{t\} - \ln^2\{t\}\right)} \tag{2.18}$$

Figure 2.2 shows the plot of Jain's fairness index in the equation above as a function of k. We see that Jain's fairness index is a monotonically decreasing function of k. Therefore, replacing k with m in the equation above, equating it with α, and solving it numerically, we get a unique value of the upper bound m. Due to the monotonically decreasing nature of Jain's fairness index, as long as the value of k is less than or equal to m, the fairness achieved by FRIP will be greater than or equal to α. □

Next, in Theorem 2.2, we obtain a closed form expression for the expected identification time.

Theorem 2.2. *Let k be the ratio of number of unidentified tags in any given frame to the size of that frame. The expected time that FRIP takes to identify all tags in the population of size t is given by the following equation:*

$$E[\mathscr{T}] = T_s + \frac{\{kT_s + (e^k - k - 1)T_c + T_e\}\{t-1\}}{k} \tag{2.19}$$

Proof. Applying expectation operator to both sides of (2.3), we get

$$E[\mathscr{T}] = T_s + E\left[\sum_{i=1}^{n}(s_i \times T_s + c_i \times T_c + e_i \times T_e)\right]$$

$$= T_s + \sum_{i=1}^{n}(E[S_i] \times T_s + E[C_i] \times T_c + E[E_i] \times T_e) \tag{2.20}$$

The extra T_s in the equation above caters for the last single slot frame, as discussed in Section 2.3.4. Substituting the values of $E[S_i]$, $E[C_i]$, and $E[E_i]$ from (2.4) to (2.6) into the equation above, we get

$$E[\mathcal{T}] = T_s + \sum_{i=1}^{n} \left((kf_ie^{-k}) \times T_s + (f_ie^{-k}) \times T_e \right.$$

$$+(f_i - kf_ie^{-k} - f_ie^{-k}) \times T_c) \tag{2.21}$$

Recall that $r = 1 - e^{-k}$. Substituting f_i by t_i/k and in turn t_i by tr^{i-1} as per (2.10), we get

$$E[\mathcal{T}] = T_s + t\left\{ T_se^{-k} + T_c\frac{1 - ke^{-k} - e^{-k}}{k} + T_e\frac{e^{-k}}{k} \right\} \sum_{i=1}^{n} r^{i-1} \tag{2.22}$$

In the equation above, $\sum_{i=1}^{n} r^{i-1}$ is sum of a standard geometric series, which is equal to $(1 - r^n)/(1 - r)$. Using $r^n = 1/t$ from (2.17), we get

$$\sum_{i=1}^{n} r^{i-1} = \frac{1 - 1/t}{1 - (1 - e^{-k})} = e^k\left\{ 1 - \frac{1}{t} \right\} \tag{2.23}$$

Substituting the value of $\sum_{i=1}^{n} r^{i-1}$ from the equation above into (2.22) results in (2.19). $\qquad\square$

Now we finally define the constraint optimization problem as: calculate the optimal value of k that minimizes the expected identification time, $E[\mathcal{T}]$, given by (2.19) under the constraint $k \leq m$, where m is obtained from the numerical solution of (2.11).

We use Karush–Kuhn–Tucker (KKT) conditions to solve this optimization problem. To apply KKT conditions, we need an objective function that needs to be optimized and an inequality constraint. The objective function in this case is $E[\mathcal{T}]$, given by (2.19), and the inequality constraint is $k - m \leq 0$. Equations (2.24)–(2.27) give the four KKT conditions that must be satisfied to obtain the optimal solution.

$$\nabla E[\mathcal{T}] + \mu \nabla(k - m) = 0, \quad \text{Stationarity} \tag{2.24}$$

$$\mu(k - m) = 0, \quad \text{Slackness} \tag{2.25}$$

$$k - m \leq 0, \quad \text{Primal feasibility} \tag{2.26}$$

$$\mu \geq 0, \quad \text{Dual feasibility} \tag{2.27}$$

where μ is called KKT multiplier.

From stationarity condition in (2.24), we get

$$\frac{d}{dk}\left[T_s + \frac{\{kT_s + (e^k - k - 1)T_c + T_e\}\{t - 1\}}{k} \right]$$

$$+\mu\frac{d}{dk}(k - m) = 0$$

$$\Rightarrow \mu = -\left[\frac{\{ke^kT_c - e^kT_c - T_e + T_c\}\{t - 1\}}{k^2} \right] \tag{2.28}$$

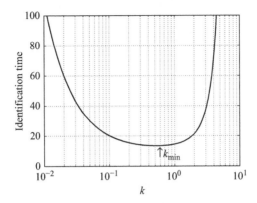

Figure 2.3   E[𝒯] vs. k

From the slackness condition in (2.25), we see that either $\mu = 0$, or $k = m$. The optimal value of k is m if the value of μ evaluated from (2.28) by using $k = m$ is not less than 0, i.e., the dual feasibility condition of (2.27) is satisfied. If, however, the dual feasibility condition is not satisfied, i.e., the value of μ evaluated from (2.28) by using $k = m$ is less than 0, then m is not the optimal value of k. In this case, we obtain the optimal value of k by putting $\mu = 0$ in (2.28) and numerically solving it for k.

To get an intuitive insight into the process of finding the solution of this constraint optimization problem as described above, consider Figure 2.3 that plots (2.19) as a function of k. Note from the figure that the identification time is a convex function of k, and thus has a minima at $k = k_{min}$. When the value of m calculated using (2.11) is less than or equal to k_{min}, then m is the optimal value of k because although the expected identification time could be reduced further by using a value of k in the range $m \leq k \leq k_{min}$, the value of fairness achieved in that case will not satisfy the requirement of $\mathscr{J} \geq \alpha$ because $\mathscr{J}$ is monotonically decreasing function of k as seen in Figure 2.2. Relating this to KKT conditions, when $m \leq k_{min}$, then using $k = m$ in (2.28) results in $\mu \geq 0$, i.e., the dual feasibility condition is satisfied. Thus, in this case m is indeed the optimal value of k. If, however, the value of m calculated using (2.11) is greater than k_{min}, then although using $k = m$ will still satisfy the requirement $\mathscr{J} \geq \alpha$, the expected identification time will not be minimum. Therefore, instead of using $k = m$, we should use $k = k_{min}$, because using $k = k_{min}$ will result in smallest expected identification time and at the same time the requirement $\mathscr{J} \geq \alpha$ will still be satisfied because $\mathscr{J}$ is monotonically decreasing function of k. Relating this to KKT conditions, when $m > k_{min}$, then using $k = m$ in (2.28) results in $\mu < 0$. Thus, in this case m is not the optimal value of k. The optimal value is obtained by using $\mu = 0$ in (2.28) and solving it for k. The optimal value of k obtained this way is equal to k_{min}. Figure 2.4 shows that the expected identification time per tag, calculated by (2.19) with optimal k, increases with the increase in the required fairness, i.e., *FRIP optimally trades-off the identification time to achieve the required fairness.*

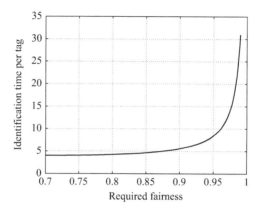

Figure 2.4 $E[\mathcal{T}]/t$ vs. α

2.3.6   Large frame size implementation

For large tag populations, the optimal frame size can exceed 2^{15}, which is not supported by the C1G2 compliant RFID tags and readers. Next, we present a strategy to execute frames with such sizes by using the SELECT command, standardized in the C1G2 standard. Using the SELECT command, a reader can broadcast a memory range and a bit mask to specify which tags should participate in the identification process. Each tag compares the bit mask with the bits in the specified range in its memory and participates in identification process only if the bit mask matches the specified bits in its memory. Using SELECT command, FRIP first divides the population into smaller groups of roughly equal sizes and then identifies the tags in each group independently using frames of allowed sizes.

To divide the whole population into groups of equal sizes, we leverage the fact that in large populations, the expected number of tags whose IDs have the least significant bit (LSB) of 0 is approximately the same as the expected number of tags whose IDs have the LSB of 1. Similarly, the expected number of tags whose IDs have the two LSBs of 00 is approximately the same as the expected number of tags whose IDs have the two LSBs of 01, 10, or 11, and so on. Therefore, a reader can divide the tag population into 2^z groups of roughly equal sizes by specifying appropriate masks for the memory range corresponding to the z LSBs of tag IDs.

To execute ith frame, whose size, f_i, is greater than $f_{max} = 2^{15}$, the reader first determines the number of groups, z, that it should divide the tag population into by calculating $\lceil \log_2 \{f_i/f_{max}\} \rceil$. It then executes 2^z *short* frames of size $\lceil f_i/2^z \rceil$ each instead of executing one *long* frame of size f_i. A short frame is a frame in which SELECT command is used and a long frame is a frame in which it is not. In each of the 2^z short frames, the reader uses SELECT command to specify a unique bit mask containing z bits for the z LSBs of tag IDs. In each short frame, on average, $t_i/2^z$ tags participate, where t_i is the number of unidentified tags at ith frame's start. Using the SELECT command in this way ensures that any given tag participates in exactly one short frame. Therefore, FRIP counts these 2^z short frames as one long frame.

At the end of these 2^z short frames, FRIP performs reestimation as described in Section 2.1.3 and continues with the $i + 1$st long frame.

Theorem 2.3 shows that executing 2^z short frames achieves same fairness as one long frame.

Theorem 2.3. *Let the optimal frame size for the ith long frame be f_i, where $f_i > f_{max}$. Let the number of unidentified tags at the start of this frame be t_i. The fairness achieved by executing 2^z short frames of size $\frac{f_i}{2^z}$ is the same as the fairness achieved by executing one long frame of size f_i.*

Proof. The value of fairness depends on the number of long frames, n, and number of singleton slots, s_i, in any long frame i, as seen in (2.2). The value of n does not change because 2^z short frames represent one long frame. Next we show that the expected value of s_i obtained using 2^z short frames is the same as the expected value of s_i obtained using one long frame, which is $s_i = kf_ie^{-k}$. When 2^z short frames are executed instead of one long frame, let we represent the number of singleton slots in the jth short frame with s_{ij}, where $1 \leq j \leq 2^z$. With the use of SELECT command as described earlier, the average number of tags that participate in the jth short frame is $t_i/2^z$. As the size of each short frame is, $f_i/2^z$, the value of ratio of number of unidentified tags that participate in each short frame to the size of frame is still k, i.e., $\frac{t_i/2^z}{f_i/2^z} = \frac{t_i}{f_i} = k$. Therefore, according to (2.4), $s_{ij} = k\frac{f_i}{2^z}e^{-k}$. As $s_i = \sum_{j=1}^{2^z} s_{ij}$, therefore, $s_i = \sum_{j=1}^{2^z} k\frac{f_i}{2^z}e^{-k} = 2^z \times k\frac{f_i}{2^z}e^{-k} = kf_ie^{-k}$. Thus, fairness remains the same. □

2.4 Experimental results

We implemented and compared the performance of FRIP with nine prior tag identification protocols, namely the three nondeterministic (Aloha [30], BS [4], and ABS [16]), the four deterministic (TH [44], TW [14], ATW [26], and STT [21]), and the two hybrid protocols (MAS [17] and ASAP [22]). We did not compare FRIP with FNFRA [46] because the two protocols address different problems: FNFRA focuses on making reader scheduling process fair while FRIP focuses on making the tag identification process fair. Although FRIP is compatible with existing reader scheduling protocols, if we implement FRIP in conjunction with an existing reader scheduling protocol and compare its performance with FNFRA, we will essentially be comparing the fairness of the existing reader scheduling protocol with FNFRA, not FRIP.

Next, we first evaluate the fairness and identification time of FRIP and study how they are affected if the tag population is split because optimal frame size is greater than 2^{15}. Then we present results from our side-by-side comparison of FRIP with existing protocols in terms of four metrics: (1) fairness, (2) number of tag transmissions, (3) energy consumption, and (4) identification time. In our simulations, we choose tag ID length to be the C1G2 standard 64 bits. We performed our simulations on two different types of populations: tag populations with IDs distributed uniformly in the ID space and tag populations with IDs distributed nonuniformly in the ID space.

For the uniform case, we range tag population sizes from 100 tags to 100,000 tags. For the nonuniform case, we distribute tag population, containing 5000 tags, in blocks, where each block is a continuous sequence of tag IDs. We range block sizes from 5 to 1,000. Our motivation for simulations on populations with tag IDs in blocks is that in some applications, such as supply chains, tag IDs often come in such blocks when they are manufactured. For each tag population size and block size, we run each protocol 10,000 times and then calculate the values of the four metrics. For example, we see how many times each tag transmits in 10,000 rounds of identification for a given protocol and then calculate the value of fairness. We do this to simulate the execution of identification protocols over a period of several days instead of a single identification round. In all the results reported for FRIP, we have incorporated the overhead of the estimation protocol that FRIP uses to obtain the initial estimate of tag population size.

2.4.1 Evaluation of FRIP

We evaluated FRIP for four different values of required fairness: $\alpha = 0.80, 0.90, 0.95$, and 0.99.

2.4.1.1 Fairness

Our results show that FRIP always achieves the required fairness irrespective of the size and type of the tag population. Figure 2.5(a) and 2.5(b) shows the fairness achieved by FRIP in the number of transmissions per tag in uniform and non-uniform tag populations. The dashed horizontal lines in these two figures show the required fairness and the solid lines with markers show the actual fairness achieved by FRIP. We see that the solid lines always lie on or above the dashed lines, which shows that FRIP always achieves the required fairness.

2.4.1.2 Identification time

Our results show that the identification time of FRIP obtained from simulations matches the theoretical minimum time obtained from (2.19) using the optimal value

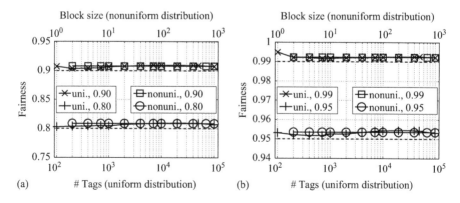

Figure 2.5   Fairness of FRIP: (a) $\alpha = 0.80, 0.90$ and (b) $\alpha = 0.95, 0.99$

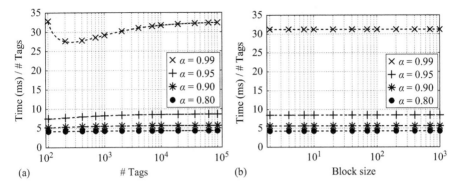

Figure 2.6   Identification Time of FRIP: (a) uniform distribution and (b) nonuniform distribution

of k, for all tag population sizes and distributions. Figure 2.6(a) shows the identification time of FRIP per tag for uniformly distributed tag populations using four different values of required fairness. Similarly, Figure 2.6(b) shows the identification time of FRIP per tag for nonuniformly distributed tag populations using four different values of required fairness. The durations of singleton, collision, and empty slots vary among the readers and are typically around 3, 1.5, and 0.3 ms, respectively, as reported in [44]. We have used these values of 3, 1.5, and 0.3 ms as durations of singleton, collision, and empty slots in our simulations. The dashed lines in these two figures represent the theoretical minimum identification times calculated from (2.19) using the optimal values of k for corresponding values of α. The markers represent the average normalized identification times of FRIP obtained from simulations. We see that the markers always lie on their respective dashed lines, which shows that the identification time of FRIP obtained from simulations matches exactly with the minimum value of the identification time of FRIP calculated theoretically. This shows the correctness of the analytical model and that the independence assumption described in Section 2.3.3 does not create any problems. FRIP requires just 7.5 ms of identification time per tag to achieve a fairness of 0.95. Although the identification time increases with the increase in required fairness, it is still very small. For example, FRIP takes only one and a quarter minute to identify 10,000 tags when $\alpha = 0.95$.

2.4.1.3   Effect of splitting tag population

Splitting the tag population due to optimal frame size greater than 2^{15} does not affect the fairness and identification time of FRIP. Figure 2.7(a) shows the fairness achieved by FRIP in the number of transmissions per tag when tag populations are split into smaller groups and when they are not split. The dashed horizontal lines in this figure show the required fairness and the solid lines with markers show the actual fairness achieved. We see that the solid lines always lie above the dashed lines for both cases i.e., when tag populations are not split and frame sizes greater than 2^{15} are used and when tag populations are split and frame sizes smaller than 2^{15} are used, as

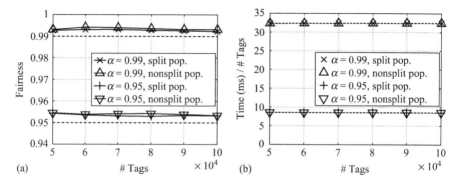

Figure 2.7   *Effect of splitting the tag population: (a) fairness and*
(b) identification time

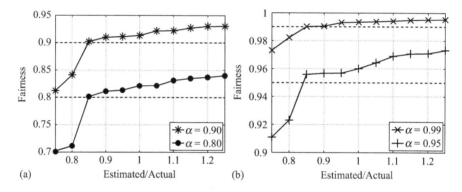

Figure 2.8   *Effect of error in population size estimate: (a) $\alpha = 0.80, 0.90$ and*
(b) $\alpha = 0.95, 0.99$

described in Section 2.3.6. Figure 2.7(b) shows the normalized identification time of FRIP for split as well as nonsplit tag populations. The dashed lines in this figure represent the theoretical minimum identification times calculated from (2.19) using the optimal values of k for corresponding values of α. The markers represent the average normalized identification times of FRIP with split and nonsplit populations. We see that the markers for split and nonsplit populations overlap and always lie on their respective dashed lines.

2.4.1.4   Effect of error in population size estimate

FRIP tolerates an error of up to 15% in population size estimate before its fairness drops below the required fairness. Figure 2.8(a) and 2.8(b) plots the fairness achieved by FRIP for different values of errors in population size estimates. The horizontal axes in these figures represent the ratio of estimated tag population size to actual tag population size. The dashed horizontal lines show the required fairness and the solid lines with markers show the actual fairness achieved. We see that the solid lines always cross the dashed lines at or before the ratio of 0.85. We also observe that as the estimated tag population size gets larger than the actual tag population size, FRIP achieves better fairness than required because it uses larger frame sizes. This means that the improved fairness comes at the cost of slow identification. Therefore, it is important that the estimate of tag population size is close to its actual size. In our simulations, we configure ART such that the estimates lie within ±5% of the actual value.

2.4.2 Comparison with existing protocols

We compare FRIP with existing protocols in terms of following four metrics: fairness, normalized number of tag transmissions, normalized power consumption, and normalized identification time. The term *normalized* means that we report these metrics on per tag basis, i.e., we first calculate the value of the metric for the whole population and then divide it by the size of the tag population. Compared to nonnormalized values, normalized values of metrics give a better idea of the effect of population size on the metric. For example, the nonnormalized value of the identification time for the whole population is bound to increase with the increase in tag population size, and thus, does not indicate whether the identification time per tag stays constant or changes with the change in population size.

2.4.2.1 Fairness

Our results show that FRIP achieves higher fairness than all existing protocols. This is because FRIP takes required fairness as input and can achieve arbitrarily high fairness (such as 0.99) by optimally trading-off identification time for it. In contrast, the existing protocols do not take the required amount of fairness as input. Figure 2.9(a) and 2.9(b) shows the fairness achieved by FRIP and the nine existing protocols for populations with uniform and nonuniform distributions, respectively. Among the existing protocols, TH achieves the highest fairness, which is bounded above by 0.84.

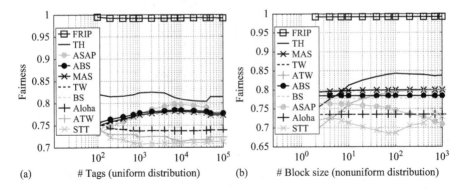

Figure 2.9 Fairness comparison: (a) uniform distribution and (b) nonuniform distribution

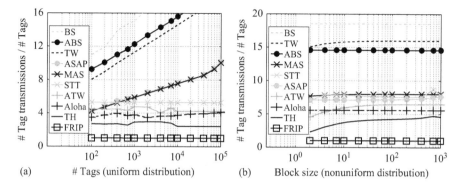

Figure 2.10   Normalized transmissions comparison: (a) uniform distribution and (b) nonuniform distribution

2.4.2.2   Normalized tag transmissions

Our results show that the average number of transmissions per tag are the fewest for FRIP compared to existing protocols. For example, when FRIP is configured for a required fairness of 0.99, it reduces the normalized tag transmissions compared to the best prior C1G2 compliant nondeterministic, deterministic, and hybrid tag identification protocols by an average of 3.74, 2.62, and 6.85 times, respectively, for uniformly distributed tag populations, and by an average of 5.50, 3.88, and 7.86 times, respectively, for nonuniformly distributed tag populations. Fewer transmissions per tag means lesser power consumption at the tags, which increases the battery life. Figure 2.10(a) and 2.10(b) shows the normalized tag transmissions of all protocols for uniformly and nonuniformly distributed tag populations, respectively.

2.4.2.3   Normalized energy consumption

Our results show that the tags consume least amount of energy when the reader executes FRIP. Although we have already shown that the number of transmissions per tag are the fewest for FRIP, this does not guarantee that the overall power consumed per tag is also the smallest, because empty slots also consume some power. A tag consumes approximately 1 mW of power during singleton slots and 35 mW of power during empty and collision slots [17]. Using these values of power, when FRIP is configured for a required fairness of 0.99, it reduces the power consumption per tag compared to the best prior C1G2 compliant nondeterministic, deterministic, and hybrid tag identification protocols by an average of 15.2%, 11%, and 28.7%, respectively, for uniformly distributed tag populations, and by an average of 14.4%, 11.3%, and 384%, respectively, for non-uniformly distributed tag populations. Figure 2.11(a) and 2.11(b) shows the normalized power consumption of all protocols for uniformly and nonuniformly distributed tag populations, respectively. FRIP consumes the least amount of power while achieving the highest fairness.

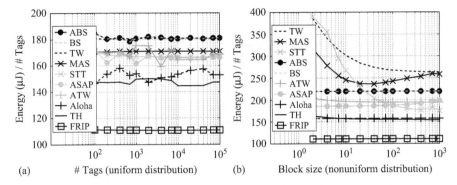

Figure 2.11 *Normalized energy comparison: (a) uniform distribution and (b)*
nonuniform distribution

Table 2.2 *Per tag identification time comparison*

Protocol	Fairness	Protocol (ms/Tag)	FRIP (ms/Tag)	Improvement (%)
STT	0.71	4.64	4.05	14.57
ATW	0.72	4.37	4.05	07.90
Aloha	0.74	4.93	4.07	21.13
ABS	0.78	5.30	4.17	27.10
BS	0.78	5.30	4.17	27.10
MAS	0.78	4.72	4.17	13.19
TW	0.78	5.30	4.17	27.10
ASAP	0.79	4.59	4.22	08.77
TH	0.82	3.78	4.39	−13.90

2.4.2.4 Normalized identification time

Our results show that identification time of FRIP is smaller than all existing protocols,
except one, when the reader executes FRIP with the required fairness equal to what
the corresponding existing protocols achieve. As the identification time of FRIP is a
function of required fairness, for a fair comparison, we first determined the fairness
value that each protocol achieves, and then executed FRIP using those values as the
required fairness. Table 2.2 reports the fairness and identification times per tag for
all existing protocols in the second and third columns, respectively. It also reports
the identification times per tag of FRIP in the fourth column when executed for the
required fairness values in the second column. FRIP is 21.13% and 13.19% faster
than the best prior C1G2 compliant nondeterministic and hybrid tag identification
protocols, respectively. FRIP is faster than all deterministic protocols as well except
TH, which is 13.9% faster than FRIP. Although TH is faster, it cannot achieve an
arbitrarily high fairness.

2.5   Conclusion

The key technical contribution of this chapter is in proposing a fair identification protocol for reading the IDs of all tags in an RFID tag population such that the number of transmissions per tag are equal. The key technical depth of this chapter is in the mathematical development of the theory that FRIP is based upon. The solid theoretical underpinning of FRIP ensures that the actual fairness that it achieves is greater than or equal to the required fairness. We have developed a novel technique that FRIP uses to execute large frame sizes to ensure compliance with the C1G2 standard. We have presented a comprehensive evaluation of FRIP and its side-by-side comparisons with nine major prior tag identification protocols. Our experimental results show that FRIP significantly outperforms all prior identification protocols, even those that are not C1G2 compliant, for not only the fairness but also the metrics such as number of transmissions per tag and power consumption.

Chapter 3
RFID estimation—design and optimization

3.1 Introduction

3.1.1 Motivation and problem statement

Radio frequency identification (RFID) systems are widely used in various applications such as object tracking [19], 3D positioning [29], indoor localization [20], supply chain management [15], inventory control, and access control [8,18] because the cost of commercial RFID tags is negligible compared to the value of the products to which they are attached (e.g., as low as 5 cents per tag [24]). An RFID system consists of tags and readers. A tag is a microchip with an antenna in a compact package that has limited computing power and communication range. There are two types of tags: (1) passive tags, which are powered up by harvesting the radio frequency energy from readers (as they do not have their own power sources) and have communication range often less than 20 ft; (2) active tags, which have their own power sources and have relatively longer communication range. A reader has a dedicated power source with significant computing power. It transmits a query to a set of tags and the tags respond over a shared wireless medium.

This chapter concerns the fundamental problem of estimating the size of a given tag population. This is needed in many applications such as tag identification, privacy sensitive RFID systems, and warehouse monitoring. In tag identification protocols, which read the ID stored in each tag, population size is estimated at the start to guide the identification process [44]. For example, for tag identification protocols that are based on the framed-slotted Aloha protocol [standardized in EPCGlobal Class-1 Generation-2 (C1G2) RFID standard [53] and implemented in commercial RFID systems], tag estimation is often used to calculate the optimal frame size. In privacy sensitive RFID systems, such as those used in parks for continuously monitoring the number of visitors in different areas of a park to plan the guided trips efficiently, readers may not have the permission to identify human individuals. In warehouses with RFID-based monitoring systems, managers often need a quick estimate of the number of products left in stock for various purposes such as the detection of employee theft. Note that although tag population size can be accurately measured by tag identification, the speed will be too slow.

We formally define the tag estimation problem as: *given a tag population of unknown size t, a confidence interval β ∈ (0, 1], and a required reliability α ∈ [0, 1),* a set of readers needs to collaboratively compute the estimated number of tags $\tilde{t}$ so that $P\left\{|\tilde{t} - t| \le \beta t\right\} \ge \alpha$. When the number of readers is one, we call this problem *single-reader estimation*; otherwise, we call this problem *multireader estimation*. A tag estimation scheme should satisfy the following three requirements:

1. *Reliability*: The actual reliability should always be greater than or equal to the required reliability. The reliability α given as input is called the *required reliability*. The reliability that an estimation scheme achieves is called its *actual reliability*.
2. *Scalability*: The estimation time needs to be scalable to large population sizes because in many applications, the number of passive tags can be very large due to their low cost, easy disposability, and powerless operation.
3. *Deployability*: The estimation scheme needs to be compliant with the C1G2 standard and should not require any changes to tags.

3.1.2   Proposed approach

In this chapter, we propose a new scheme called *Average Run-based Tag estimation* (*ART*), which satisfies all of the above three requirements. The communication protocol used by ART is the standardized framed-slotted Aloha protocol, in which a reader first broadcasts a value f to the tags in its vicinity where f represents the number of time slots present in a forthcoming frame. Then each tag randomly picks a time slot in the frame and replies during that slot. Thus, the reader gets a binary sequence of 0s and 1s by representing a slot with no tag replies as 0 and a slot with one or more tag replies as 1. The key idea of ART is to estimate tag population size based on the average run size of 1s in the binary sequence. We show that the average run size of 1s in a frame monotonously increases with the increase in the size of tag population. Thus, average run size of 1s is an indicator of tag population size.

3.1.3   Advantages of ART over prior art

ART is advantageous in terms of speed and deployability. For speed, ART is faster than all prior schemes. For example, given a confidence interval of 0.1% and the required reliability of 99.9%, ART is consistently seven times faster than the fastest existing schemes (i.e., UPE [50] and EZB [54]) for any tag population size. The reason behind ART being faster than prior schemes is that the new estimator that we propose in this chapter, namely the average run size of 1s, has significantly smaller variance compared to the estimators used in prior schemes (such as the total number of 0s [50,54] and the location of the first 1 in the binary sequence [11]), as we analytically show in Section 3.7.3. An estimator with small variance is faster because the Aloha frames need to be repeated fewer times to achieve the required reliability. Furthermore, the estimation time of ART is provably independent of tag population sizes. In contrast, as tag volume increases, the estimation time of some prior schemes (e.g., FNEB [11]) increases.

For deployability, ART neither requires modification to the tags nor to the communication protocol between tags and readers. ART only needs to be implemented on the reader side as a software module without any hardware modifications. ART also does not demand any unpractical system parameters beyond the C1G2 standard. In contrast, some prior schemes require modification to tags and some demand unrealistic system parameters. For example, the scheme in [23] requires each tag to store thousands of hash functions, which is not practical to implement on passive tags and is not compliant with the C1G2 standard. As another example, the scheme in [11] uses increasingly large frame sizes as population size increases (e.g., the frame size required by the scheme in [11] is greater than half of tag population size), which soon exceeds the maximum limit allowed by the C1G2 Standard.

Chapter organization
Section 3.2 describes the prominent related work in RFID estimation. Section 3.3 gives an overview of how ART works. It also describes the assumptions in the formal development of ART and their implications. Section 3.4 gives the detailed formal development of ART. Section 3.5 derives mathematical expressions to calculate the optimal values of system parameters, which ensure that ART achieves the required reliability in smallest possible time. Section 3.6 describes how ART estimates sizes of arbitrarily large tag populations and how it handles environments with multiple RFID readers. Section 3.7 gives the proof that estimation time of ART is independent of tag population size and discusses its computational complexity. It also presents analytical comparison of ART with existing schemes to mathematically justify the faster speed of ART compared to existing schemes. Section 3.8 presents results from our experimental evaluation of ART along with side-by-side comparisons with existing schemes. Finally, we conclude the chapter in Section 3.9.

3.2 Related work

The first tag estimation scheme, called Unified Probabilistic Estimator (UPE), was proposed by Kodialam and Nandagopal in 2006 [50]. UPE uses the framed-slotted Aloha protocol and makes estimation based on either the number of empty slots or that of collision slots in a frame. Besides this estimator having larger variance than ART, UPE requires the differentiation among empty, single, and collision slots, which takes significantly more time than differentiating between empty and non-empty slots. According to C1G2, a reader requires 300 μs to detect an empty slot, 1500 μs to detect a collision, and 3000 μs to complete a successful read. In [54], Kodialam *et al.* proposed an improved framed-slotted Aloha protocol based estimation scheme called Enhanced Zero Based (EZB) estimator, which performs estimation based on the total number of 0s in a frame. While UPE estimates population size in each round and averages the estimated sizes when all rounds are finished, EZB only records the total number of 0s in each frame and at the end of all rounds, EZB first averages the recorded values and then uses it to do estimation.

In [23], Qian *et al.* proposed an estimation scheme called Lottery Frame (LoF). Compared to UPE and EZB, LoF is faster; however, it is impractical to implement as it requires each tag to store a large number (i.e., the number of bits in a tag ID times the number of frames, which can be in the scale of thousands) of unique hash functions. LoF needs to modify both tags and the communication protocol between readers and tags, which makes it non-compliant with C1G2. Han *et al.* proposed a tag estimation scheme called First Non Empty Based (FNEB) estimator, which is based on the size of the first run of 0s in a frame [11]. FNEB is based on an assumption that frame size can be arbitrarily large, which does not hold in practice. Li *et al.* proposed an estimation scheme called Maximum Likelihood Estimator (MLE) for active tags with the goal of minimizing power consumption of active tags [75]. In [57], Shah and Wong proposed a multireader tag estimation scheme which is based on an unrealistic assumption that any tag covered by multiple readers only replies to one reader. In [58], Zanella proposed Collision Set Estimator (CSE) that utilizes maximum likelihood estimation to estimate the number of tags in a population. CSE does not take accuracy requirements (α and β) as input and, therefore, cannot achieve any arbitrary required reliability.

3.3 ART—scheme overview

3.3.1 Communication protocol overview

ART uses the framed-slotted Aloha protocol specified in C1G2 as its MAC layer communication protocol. In this protocol, the reader first tells tags the frame size f and a random seed number R. Later in the chapter, we will see how a simple use of seed number R will make it straightforward to extend our estimation scheme to use multiple readers with overlapping regions. Each tag within the transmission range of the reader then uses f, R, and its *ID* to select a slot in the frame by evaluating a hash function $h(f, R, ID)$ whose result is in $[1, f]$ following a uniform distribution. Each tag has a counter initialized with the slot number it chose to reply. After each slot, the reader first transmits an end of slot signal and then each tag decrements its counter by one. In any given slot, all the tags whose counters are equal to 1 respond to the reader. In essence, each tag picks a random slot from 1 to f following a uniform distribution. If no tag replies in a slot, it is called an *empty slot*; if exactly one tag replies, it is called a *singleton slot*; and if two or more tags reply, it is called a *collision slot*.

3.3.2 Estimation scheme overview

At the end of a frame, the reader obtains a sequence of 0s and 1s by representing an empty slot with 0 and a singleton or collision slot with 1. In this binary sequence, a *run* is a subsequence where all bits in this subsequence are 0s (or 1s) but the bits before and after the subsequence are 1s (or 0s), if they exist. For example, 011100 has 3 runs: 0, 111, and 00.

ART uses the average run size of 1s to estimate tag population size. The intuition is that as tag population size increases, the average run size of 1s increases (and that of 0s decreases). We illustrate this intuition using the simulation results in Figure 3.1, which shows that the average run size of 1s increases as tag population size increases from 0 to 160. The markers in this figure are the average of 100 runs. The lines above

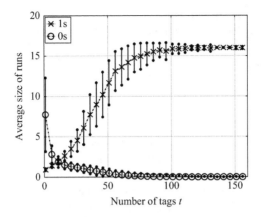

Figure 3.1   Average run size of 0s and 1s vs. tag population size t ($f = 16$)

and below each marker show the standard deviation of the experiments. This figure shows that given a tag population size and a frame size, there is a distinct expected value of the average run size of 1s. The expected value of the average run size of 1s is a monotonic function of the number of tags, which means that a unique inverse of this function exists. Thus, given the observed average run size of 1s, using the inverse function, we can get the estimated value $\tilde{t}$ of tag population size t. Similar to other tag estimation schemes, ART also uses multiple frames obtained from multiple rounds of the framed-slotted Aloha protocol to reduce its estimation variance and therefore increase its estimation reliability. Using different seed values for different frames, in each frame, the same tag will choose a different slot to respond.

To scale to large tag population sizes, ART uses a persistence probability p by which a tag decides whether it should reply to the reader in a given frame. The persistence probability was first introduced in [50]. To avoid making any modification to tags, this probability is implemented by "virtually" extending frame size $1/p$ times, i.e., the reader announces a frame size of f/p but terminates the frame after the first f slots. According to C1G2, the reader can terminate a frame at any point. By adjusting p, ART is able to estimate tag populations of large sizes.

3.3.3   Formal development: overview and assumptions

To formally develop an estimator, we first need to derive the equation for the expected value of average run size of 1s as a function of frame size f, tag population size t, and persistence probability p. We then use the inverse of this function to get the estimated value $\tilde{t}$ from the observed value of the average run size of 1s. To achieve the required reliability in minimum estimation time, we optimize f, p, and the number of rounds n so that the total number of slots $(f + l) \times n$ is minimized while satisfying $P\{|\tilde{t} - t| \le \beta t\} \ge \alpha$. Here, l is a constant that represents the C1G2 specified mandatory time delay in terms of number of empty slots between the end of a frame and the start of next frame. Typically, this delay is about 1 ms (i.e., $l \approx 3.33$ empty slots) [53,56].

To make the formal development tractable, we assume that instead of picking a single slot to reply at the start of frame of size f, a tag independently decides to reply in each slot of the frame with probability $1/f$ regardless of its decision about previous or forthcoming slots. Vogt first used this assumption for the analysis of framed-slotted Aloha protocol for RFID and justified its use by recognizing that this problem belongs to a class of problems known as "occupancy problems," which deals with the allocation of balls to urns [49]. Ever since, the use of this assumption has been a norm in the formal analysis of all Aloha-based RFID protocols [11,23,30,45,49,50,52,54,57,75].

The implication of this assumption is that when a tag independently chooses a slot to reply, it can end up choosing more than one slots in the same frame or even not choosing any at all, which is not in accordance with C1G2 standard that requires a tag to pick exactly one slot in a frame. However, even with the independence assumption, the expected number of slots that a tag chooses in a frame is still one. As we draw our estimate from a large number of frames to achieve required reliability, we can expect to observe this expected number. Therefore, the analysis with the assumption of independence is asymptotically the same as that without the independence assumption. Bordenave *et al.* further explained in detail why this independence assumption in analyzing Aloha-based protocols provides results just as accurate as if all the analysis was done without this assumption [51]. Note that this independence assumption is made only to make the formal development tractable. In all the simulations we have presented in this chapter, a tag chooses exactly one slot at the start of frame.

3.4   ART—estimation algorithm

Next, we first focus on the single-reader version of ART. In Section 3.6.2, we will present a method to extend ART to handle multiple-readers with overlapping regions. Table 3.1 lists the symbols used in this chapter.

For ART, in each round of the Aloha protocol, we calculate the average run size of b. For example, the average run size of 1 in frame 01110011 (which has two runs of 1, i.e., 111 and 11) is $(3 + 2)/2 = 2.5$. After n rounds, we obtain n average run sizes of b and then calculate the average of these n values. This final value is then substituted for the expected value of the average run size of b in a frame to estimate the tag population size.

The probability that a slot in a frame is b, where $b = 0$ or 1, can be calculated using Lemma 3.1.

Lemma 3.1. *Let t be the actual tag population size, f be the frame size, p be the persistence probability (i.e., the probability that a tag participates in a frame), and q_b be the probability that a slot in a frame is b. Thus:*

$$
q_b = \begin{cases} (1 - \frac{p}{f})^t & \text{if } b = 0 \\ 1 - (1 - \frac{p}{f})^t & \text{if } b = 1 \end{cases} \tag{3.1}
$$

Table 3.1 Symbols used in the chapter

Symbol	Description
t	Actual tag population size
t_m	Upper bound on # of tags
L_{tm}, U_{tm}	Bounds within which t_m/t should lie
t_M	Maximum # of tags that can be estimated
$\tilde{t}$	Estimated # of tags
α	Required reliability
$\tilde{\alpha}$	Expected value of actual reliability
β	Required confidence interval
f	Frame size
f_{op}	Optimal frame size
n	# Of rounds (i.e., frames)
p	Persistence probability
R	Random seed from reader
$h(f, R, ID)$	Uniform hash function with output in $[1, f]$
b	Value of a slot: $b = 0$ or $b = 1$
$\bar{b}$	$1 - b$
q_b	Probability that a slot is b
$E[.]$	Expected value
$\text{Var}(.)$	Variance
$\text{Cov}(.)$	Covariance
Y_b	Random variable for # of b slots in frame
y	Element of sample space of Y_b
R_b	Random variable for # of runs of b in frame
r	Element of sample space of R_b
X_b	Random variable for average run size of b in a frame
q_1, q_1^+, q_1^-	Probability that a slot is 1 when number of tags in population are t, $(1 + \beta)t$, and $(1 - \beta)t$, respectively
$\mu\{.\}$	Expected value of X_b
$\sigma\{.\}$	Standard deviation of X_b
$\xi\{f, y, r\}$	Number of ways in which y occurrences of b and $f - y$ occurrences of $\bar{b}$ can be arranged in f slots while ensuring that the number of runs of b are r
η	Simplification variable: $\eta = (1 - \frac{p}{f})^t$
z	# Of bits the reader uses in the mask
l	Constant representing delay between consecutive frames in terms of number of empty slots

Proof. The probability that a tag chooses a given slot in a frame is p/f. The probability that it does not choose that slot is $1 - \frac{p}{f}$. The probability that none of the tags choose that slot is $(1 - \frac{p}{f})^t$, which is the value of q_0. As the tags choose the slots independently, q_b is the same for each slot of the frame. The probability that a slot is chosen by at least one tag is $1 - q_0$, which is the value of q_1. $\square$

Let X_b be the random variable representing the average run size of b in a frame. Next, we calculate the expectation and variance of X_b. The expectation of X_b will be used to estimate the tag population size and the variance of X_b will be used to calculate the values of p, n, and f that will ensure that the actual reliability is greater than the required reliability and the estimation time is minimum. Let Y_b be the random variable representing the number of times b occurs in a frame and R_b be the random variable representing the number of runs of b in a frame. By definition, $X_b = \frac{Y_b}{R_b}$ holds for any frame. Next, we first calculate $E[Y_b]$, $\mathrm{Var}(Y_b)$, $E[R_b]$, $\mathrm{Var}(R_b)$, and $\mathrm{Cov}(Y_b, R_b)$ in Lemmas 3.2 and 3.3. Then, we use them to calculate $E[X_b]$ and $\mathrm{Var}(X_b)$ in Theorem 3.1. Using (3.20) in Theorem 3.1, replacing $E[X_b]$ by the observed average run size of b from n frames, we obtain an equation with only one unknown t. Finally, we use Brent's method to obtain the numerical solution of this equation. The result is the estimated tag population size $\tilde{t}$. Since ART uses X_b to estimate the tag population size, we call X_b the *estimator* of ART.

Lemma 3.2. *Let Y_b be the random variable representing the number of times b occurs in a frame and R_b be the random variable representing the number of runs of b in a frame. Given tag population size t, frame size f, and persistence probability p, we have:*

$$E[Y_b] = fq_b \tag{3.2}$$

$$\mathrm{Var}(Y_b) = fq_b(1 - q_b) \tag{3.3}$$

$$E[R_b] = q_b\,(q_b + f(1 - q_b)) \tag{3.4}$$

$$\mathrm{Var}(R_b) = f(q_b - 4q_b^2 + 6q_b^3 - 3q_b^4) + (3q_b^2 - 8q_b^3 + 5q_b^4) \tag{3.5}$$

Proof. Each slot i of frame f has probability q_b of being b. Therefore, $Y_b \sim$ Binom(f, q_b). Using general formula for expectation and variance of a binomial random variable, $E[Y_b]$ and $\mathrm{Var}(Y_b)$ are given by (3.2) and (3.3).

Let γ_1, γ_2, ..., γ_f represent the sequence of binary random variables representing the value of each slot in a frame of size f. Since each tag randomly and independently picks a slot in the frame, all γ_i are identically distributed. Furthermore, $P\{\gamma_i = b\} = q_b$. Let $\bar{b} = 1 - b$ and let I_i be the indicator random variable whose value is 1 if a run of b begins at γ_i.

$$I_i = \begin{cases} 1 & \text{if } (\gamma_i = b, i = 1) \vee (\gamma_i = b \wedge \gamma_{i-1} = \bar{b}, i > 1) \\ 0 & \text{otherwise} \end{cases} \tag{3.6}$$

Thus,

$$R_b = \sum_{i=1}^{f} I_i \tag{3.7}$$

because

$$E[I_i] = \begin{cases} P\{\gamma_i = b\} = q_b & \text{if } i = 1 \\ P\{\gamma_{i-1} = \bar{b}, \gamma_i = b\} = q_b(1 - q_b) & \text{if } i > 1 \end{cases} \tag{3.8}$$

we get

$$E[R_b] = \sum_{i=1}^{f} E[I_i] = q_b + \sum_{i=2}^{f} q_b(1 - q_b) = q_b(q_b + f(1 - q_b)) \qquad (3.9)$$

As R_b is sum of f random variables, some of which are correlated, we use the general expression for variance of sum of correlated random variables to obtain the variance of R_b.

$$\text{Var}(R_b) = \text{Var}\left(\sum_{i=1}^{f} I_i\right) = \sum_{i=1}^{f} \text{Var}(I_i) + 2 \sum_{j=2}^{f} \sum_{\forall i < j} \text{Cov}(I_i, I_j) \qquad (3.10)$$

Here, we used the fact that the frame size is always greater than 1 during the estimation process whenever the information about runs is used. As $I_i \sim \text{Bernoulli}(q_b)$, its variance is that of a Bernoulli random variable given by:

$$\text{Var}(I_i) = E[I_i](1 - E[I_i]) \qquad (3.11)$$

Note that I_i and I_j are dependent on each other if and only if $i = j - 1$ because I_{j-1} and I_j cannot both be 1 in the same frame. Other than that, $\forall i < j - 1$, I_i and I_j are independent. Thus:

$$\text{Cov}(I_i, I_j) = \begin{cases} 0 & \text{if } i < j - 1 \\ -E[I_i]E[I_j] = -E[I_i]q_b(1 - q_b) \\ & \text{if } i = j - 1 \end{cases} \qquad (3.12)$$

Hence, we have:

$$\begin{aligned} \text{Var}(R_b) &= \text{Var}(I_1) + \sum_{j=2}^{f} \text{Var}(I_j) + 2\text{Cov}(I_1, I_2) \\ &\quad + 2 \sum_{j=3}^{f} \text{Cov}(I_{j-1}, I_j) \\ &= q_b(1 - q_b) + (f - 1)q_b(1 - q_b)\{1 - q_b(1 - q_b)\} \\ &\quad - 2q_b^2(1 - q_b) - 2(f - 2)q_b^2(1 - q_b)^2 \\ &= f(q_b - 4q_b^2 + 6q_b^3 - 3q_b^4) + (3q_b^2 - 8q_b^3 + 5q_b^4) \end{aligned} \qquad (3.13)$$

$\square$

Lemma 3.3. *Given tag population size t, frame size f, and persistence probability p, we have:*

$$\text{Cov}(Y_b, R_b) = \sum_{y=0}^{f} \sum_{r=0}^{\lceil \frac{f}{2} \rceil} yrq_b^y(1 - q_b)^{f-y}.\xi \{f, y, r\}$$

$$- E[Y_b]E[R_b] \qquad (3.14)$$

where

$$
\xi\{f,y,r\} = \begin{cases}
\binom{y-1}{r-1}\left[\binom{f-y-1}{r-2} + 2\binom{f-y-1}{r-1} + \binom{f-y-1}{r}\right] \\
\quad \text{if } r > 1 \wedge 0 < y < f \wedge r \leq y \wedge r \leq f - y - 1 \\[2mm]
\binom{y-1}{r-1}\left[2\binom{f-y-1}{r-1} + \binom{f-y-1}{r}\right] \\
\quad \text{if } r = 1 \wedge 0 < y < f \wedge r \leq y \wedge r \leq f - y - 1 \\[2mm]
1 \text{ if } r = 1 \wedge y = f \\[2mm]
1 \text{ if } r = 0 \wedge y = 0 \\[2mm]
0 \text{ otherwise}
\end{cases}
\tag{3.15}
$$

Proof. By definition, we have:

$$
\text{Cov}(Y_b, R_b) = \sum_{y=0}^{f} \sum_{r=0}^{f} y r P\{Y_b = y, R_b = r\} - E[Y_b]E[R_b]
\tag{3.16}
$$

Here, $P\{Y_b = y, R_b = r\}$ represents the probability that exactly y out of f slots in the frame are b and at the same time the number of runs of b is r. This probability is difficult to evaluate directly, but conditioning on Y_b simplifies the task.

$$
P\{Y_b = y, R_b = r\} = P\{R_b = r | Y_b = y\} \times P\{Y_b = y\}
\tag{3.17}
$$

As $Y_b \sim \text{Binom}(f, q_b)$, we have:

$$
P\{Y_b = y\} = \binom{f}{y} q_b^y (1 - q_b)^{f-y}
\tag{3.18}
$$

Now we calculate $P\{R_b = r | Y_b = y\}$ i.e., the probability of having r runs of b in a frame of size f given that y out of f slots are b. As tags choose the slots independently, each occurrence with r runs having y slots of b is equally likely. Therefore, we determine the total number of ways, denoted by $\xi\{f,y,r\}$, in which y occurrences of b and $f - y$ occurrences of $\bar{b}$ can be arranged such that the number of runs of b is r. We treat this as an ordered partition problem. First, we separate all the y occurrences of b from the frame and make r partitions of these y occurrences. Then, we create appropriate number of partitions of $f - y$ occurrences of $\bar{b}$ such that between consecutive partitions of b, the partitions of $\bar{b}$ can be *interleaved*. For r partitions of b, there are 4 possible partitions of $\bar{b}$.

1.   The frame starts with b and ends with b, implying that there are $r - 1$ partitions of $\bar{b}$, each interleaved between adjacent partitions of b.
2.   The frame starts with b and ends with $\bar{b}$, implying that there are r partitions of $\bar{b}$.
3.   The frame starts with $\bar{b}$ and ends with b, implying that there are r partitions of $\bar{b}$.
4.   The frame starts with $\bar{b}$ and ends with $\bar{b}$, implying that there are $r + 1$ partitions of $\bar{b}$.

We can make r partitions of y occurrences of b in $\binom{y-1}{r-1}$ ways and r partitions of $f - y$ occurrences of $\bar{b}$ in $\binom{f-y-1}{r-1}$ ways. Similarly, we can make $r + 1$ partitions of $f - y$ occurrences of $\bar{b}$ in $\binom{f-y-1}{r}$ ways and $r - 1$ partitions of $f - y$ occurrences of $\bar{b}$ in $\binom{f-y-1}{r-2}$ ways. The equation of $\xi\{f, y, r\}$ in the lemma statement follows from this discussion. The total number of ways in which y zeros can be arranged among f slots is $\binom{f}{y}$. Thus, we get:

$$P\{R_b = r | Y_b = y\} = \frac{\xi\{f, y, r\}}{\binom{f}{y}} \tag{3.19}$$

Substituting values from (3.18) and (3.19) in (3.17) and (3.16) results in (3.14). $\quad\square$

Theorem 3.1. *Given tag population size t, frame size f, and persistence probability p, we have:*

$$E[X_b] = \frac{E[Y_b]}{E[R_b]} - \frac{\mathrm{Cov}(Y_b, R_b)}{E^2[R_b]} + \frac{E[Y_b]}{E^3[R_b]} \mathrm{Var}(R_b) \tag{3.20}$$

$$\mathrm{Var}(X_b) = \frac{\mathrm{Var}(Y_b)}{E^2[R_b]} - \frac{2E[Y_b]}{E^3[R_b]} \mathrm{Cov}(Y_b, R_b) + \frac{E^2[Y_b]}{E^4[R_b]} \mathrm{Var}(R_b) \tag{3.21}$$

Proof. Let $g(Y_b, R_b) = X_b = \frac{Y_b}{R_b}$. The Taylor series expansion of g around (θ_1, θ_2) is given by:

$$g(Y_b, R_b) = \sum_{j=0}^{\infty} \left\{ \frac{1}{j!} \left[(Y_b - \theta_1) \frac{\partial}{\partial Y_b'} + (R_b - \theta_2) \frac{\partial}{\partial R_b'} \right]^j \times \right.$$

$$\left. g(Y_b', R_b') \right\}_{\substack{Y_b' = \theta_1 \\ R_b' = \theta_2}} \tag{3.22}$$

According to Bienaymé–Chebyshev inequality, we have $\theta_1 = E[Y_b]$ and $\theta_2 = E[R_b]$. Therefore, we get the following expansion of the Taylor series of $g(Y_b, R_b)$:

$$g(Y_b, R_b) = g(\theta_1, \theta_2) + \left[(Y_b - \theta_1) \frac{\partial g}{\partial Y_b} + (R_b - \theta_2) \frac{\partial g}{\partial R_b} \right]$$

$$+ \frac{1}{2} \left[(Y_b - \theta_1)^2 \frac{\partial^2 g}{\partial Y_b^2} + 2(Y_b - \theta_1)(R_b - \theta_2) \frac{\partial^2 g}{\partial Y_b \partial R_b} \right.$$

$$\left. + (R_b - \theta_2)^2 \frac{\partial^2 g}{\partial R_b^2} \right] + O(j^{-1}) \tag{3.23}$$

Taking the expectation of both sides, we get:

$$E[g(Y_b, R_b)] \approx \frac{1}{2} \left[\mathrm{Var}(Y_b) \frac{\partial^2 g}{\partial Y_b^2} + 2\mathrm{Cov}(Y_b, R_b) \frac{\partial^2 g}{\partial Y_b \partial R_b} \right.$$

$$\left. + \mathrm{Var}(R_b) \frac{\partial^2 g}{\partial R_b^2} \right] + g(\theta_1, \theta_2) \tag{3.24}$$

Evaluating the partial derivatives of g as required in (3.24), we get:

$$\frac{\partial^2 g(Y_b, R_b)}{\partial Y_b^2}\bigg|_{\substack{Y_b=\theta_1 \\ R_b=\theta_2}} = 0, \qquad \frac{\partial^2 g(Y_b, R_b)}{\partial Y_b \partial R_b}\bigg|_{\substack{Y_b=\theta_1 \\ R_b=\theta_2}} = -\frac{1}{\theta_2^2} \tag{3.25}$$

$$\frac{\partial^2 g(Y_b, R_b)}{\partial R_b^2}\bigg|_{\substack{Y_b=\theta_1 \\ R_b=\theta_2}} = 2\frac{\theta_1}{\theta_1^3} \tag{3.26}$$

Putting these values in (3.24) and using $\theta_1 = E[Y_b]$ and $\theta_2 = E[R_b]$, we get (3.20). The variance can be calculated as follows:

$$\text{Var}(g(Y_b, R_b)) = E[\{g(Y_b, R_b) - E[g(Y_b, R_b)]\}^2] \tag{3.27}$$

Considering that $E[g(Y_b, R_b)]$ is being squared in the expression above, we use first-order Taylor series expansion to get the value of $E[g(Y_b, R_b)]$ and substitute it in (3.27).

$$\begin{aligned} E[g(Y_b, R_b)] &= E\left[(Y_b - \theta_1)\frac{\partial g}{\partial Y_b} + (R_b - \theta_2)\frac{\partial g}{\partial R_b}\right] \\ &\quad + g(\theta_1, \theta_2) + O(j^{-1}) = \left[(0)\frac{\partial g}{\partial Y_b} + (0)\frac{\partial g}{\partial R_b}\right] \\ &\quad + g(\theta_1, \theta_2) + O(j^{-1}) \approx g(\theta_1, \theta_2) \end{aligned} \tag{3.28}$$

Substituting the value of $E[g(Y_b, R_b)]$ and using the first-order Taylor series expansion of $g(Y_b, R_b)$ in (3.27), we get:

$$\begin{aligned} \text{Var}(g(Y_b, R_b)) &= E\left[\left\{(Y_b - \theta_1)\frac{\partial g}{\partial Y_b} + (R_b - \theta_2)\frac{\partial g}{\partial R_b}\right\}^2\right] + O(j^{-1}) \\ &\approx \text{Var}(Y_b)\left(\frac{\partial g}{\partial Y_b}\right)^2 + 2\text{Cov}(Y_b, R_b)\frac{\partial g}{\partial Y_b}\frac{\partial g}{\partial R_b} + \text{Var}(R_b)\left(\frac{\partial g}{\partial R_b}\right)^2 \end{aligned} \tag{3.29}$$

Evaluating the partial derivatives of g as required in the equation above, we get:

$$\frac{\partial g(Y_b, R_b)}{\partial Y_b}\bigg|_{\substack{Y_b=\theta_1 \\ R_b=\theta_2}} = \frac{1}{\theta_2}, \qquad \frac{\partial g(Y_b, R_b)}{\partial R_b}\bigg|_{\substack{Y_b=\theta_1 \\ R_b=\theta_2}} = -\frac{\theta_1}{\theta_2^2} \tag{3.30}$$

Putting these values in (3.29) and using $\theta_1 = E[Y_b]$ and $\theta_2 = E[R_b]$, we get (3.21). $\qquad\square$

Figures 3.2 and 3.3 show the expectation and variance of X_1 calculated using (3.20) and (3.21), respectively, with $f = 16$ and $p = 1$. The dots in these figures represent the corresponding values obtained through 100 repetitions of simulation for each tag population size. These figures show that the values given by (3.20) and (3.21) track the simulation results very well, which serves as an experimental proof that the assumption "instead of picking a single slot to reply at the start of frame of size f, a tag independently decides to reply in each slot of the frame with probability $1/f$ regardless of its decision about previous or forthcoming slots" practically holds.

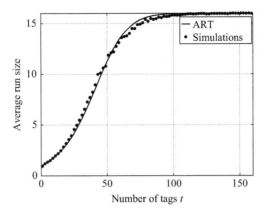

Figure 3.2   Expectation of ART estimator

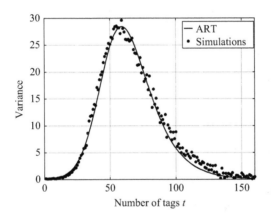

Figure 3.3   Variances of ART estimator

3.5 ART—parameter tuning

To minimize estimation time while achieving required reliability, next, we obtain values of persistence probability p, number of rounds n, and frame size f. As we have three unknowns, we require three equations that can be solved simultaneously. We derive these three equations using following three conditions: (1) the confidence interval should be symmetric around t i.e., $|\tilde{t} - t| \le \beta t$, (2) actual reliability is greater than or equal to the required reliability i.e., $P\{|\tilde{t} - t| \le \beta t\} \ge \alpha$, and (3) estimation time is minimized. We use the first condition to calculate p, the second condition to calculate n, and the last condition to calculate f.

Although both X_0 and X_1 can be used to estimate the tag population size, we choose X_1 for ART because the tag population size estimation calculated from X_1 has smaller variance compared to X_0 as we show in Section 3.7.3. It is worth noting that X_0 and X_1 are not equivalent estimators. The average run size of 0s cannot be inferred from the average run size of 1s, and vice versa. For example, 1100011 and 1100110 have the same average run size of 1s, but they have different average run size of 0s. Fundamentally, X_0 and X_1 are not equivalent estimators because for any slot, the probability of it being 0 and that of it being 1 are different.

3.5.1   Persistence probability p

We express confidence interval requirement $|\tilde{t} - t| \leq \beta t$ as:

$$(1 - \beta)t \leq \tilde{t} \leq (1 + \beta)t \tag{3.31}$$

Recall from Lemma 3.1 that we use q_1 to denote the probability that a slot in a frame is 1 when the number of tags in the population are t and the persistence probability is p. Let q_1^+ and q_1^- denote the probabilities that a slot in a frame is 1 when the number of tags in the population are $(1 + \beta)t$ and $(1 - \beta)t$, respectively, and the persistence probability is p. Let $\tilde{q}_1$ represent the estimate of q_1. Therefore, we have:

$$q_1^+ = 1 - \left(1 - \frac{p}{f}\right)^{(1+\beta)t} \Rightarrow (1 + \beta)t = \frac{\ln\{1 - q_1^+\}}{\ln\{1 - \frac{p}{f}\}} \tag{3.32}$$

$$q_1^- = 1 - \left(1 - \frac{p}{f}\right)^{(1-\beta)t} \Rightarrow (1 - \beta)t = \frac{\ln\{1 - q_1^-\}}{\ln\{1 - \frac{p}{f}\}} \tag{3.33}$$

$$\tilde{q}_1 = 1 - \left(1 - \frac{p}{f}\right)^{\tilde{t}} \quad \Rightarrow \quad \tilde{t} = \frac{\ln\{1 - \tilde{q}_1\}}{\ln\{1 - \frac{p}{f}\}} \tag{3.34}$$

Substituting values of $(1 + \beta)t$, $(1 - \beta)t$, and $\tilde{t}$ from (3.32), (3.33), and (3.34), respectively, into Expression (3.31), we get:

$$\frac{\ln\{1 - q_1^-\}}{\ln\{1 - \frac{p}{f}\}} \leq \frac{\ln\{1 - \tilde{q}_1\}}{\ln\{1 - \frac{p}{f}\}} \leq \frac{\ln\{1 - q_1^+\}}{\ln\{1 - \frac{p}{f}\}} \tag{3.35}$$

As $\ln\{1 - \frac{p}{f}\} < 0$, thus:

$$\ln\{1 - q_1^+\} \leq \ln\{1 - \tilde{q}_1\} \leq \ln\{1 - q_1^-\} \tag{3.36}$$

Exponentiating and rearranging, the confidence interval requirement becomes:

$$q_1^- \leq \tilde{q}_1 \leq q_1^+ \tag{3.37}$$

As $E[X_1]$ and $\text{Var}(X_1)$ are functions of q_1, denoting $E[X_1]$ by $\mu\{q_1\}$, $\text{Var}(X_1)$ by $\sigma^2\{q_1\}$, and the observed average value of X_1 from the n frames by $\tilde{X}_1$, we have $\tilde{q}_1 = \mu^{-1}\{\tilde{X}_1\}$. Using $\mu^{-1}\{\tilde{X}_1\}$ to substitute $\tilde{q}_1$ in the expression above, we get:

$$q_1^- \leq \mu^{-1} \tag{3.38}$$

Based on the fact that the variance of a random variable is reduced by n times if the same experiment is repeated n times, by running n rounds and getting n frames, the

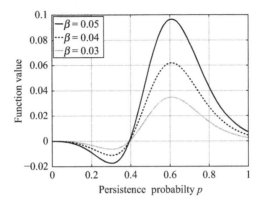

Figure 3.4   Equation (3.42) as a function of p

variance of X_1 becomes $\frac{\sigma^2\{q_1\}}{n}$ and the standard deviation of X_1 becomes $\frac{\sigma\{q_1\}}{\sqrt{n}}$. Let Z denote $\frac{\bar{X}_1 - \mu\{q_1\}}{\sigma\{q_1\}/\sqrt{n}}$. Thus, the expression above becomes

$$\frac{\mu\{q_1^-\} - \mu\{q_1\}}{\frac{\sigma\{q_1\}}{\sqrt{n}}} \leq Z \leq \frac{\mu\{q_1^+\} - \mu\{q_1\}}{\frac{\sigma\{q_1\}}{\sqrt{n}}} \tag{3.39}$$

By the central limit theorem, Z approximates a standard normal random variable. The area under the standard normal curve gives the success probability, which is the required reliability in our context. For the confidence interval to be symmetric on both the upper and lower sides of the population size as per the first of the three conditions, the absolute value of the upper and lower limits of Z should be equal. Let k represent the absolute value of these upper and lower limits. Thus, we can represent Z as follows:

$$-k \leq Z \leq k \tag{3.40}$$

From Expressions (3.39) and (3.40), we get:

$$\frac{\mu\{q_1^-\} - \mu\{q_1\}}{\frac{\sigma\{q_1\}}{\sqrt{n}}} = -k, \quad \frac{\mu\{q_1^+\} - \mu\{q_1\}}{\frac{\sigma\{q_1\}}{\sqrt{n}}} = k \tag{3.41}$$

As the absolute values of the right hand sides (R.H.S.) of both equations above are k, we get:

$$2\mu\{q_1\} - \mu\{q_1^+\} - \mu\{q_1^-\} = 0 \tag{3.42}$$

The equation above gives the condition that needs to be satisfied to make the confidence interval symmetric around the tag population size. Figure 3.4 plots the value of left hand side (L.H.S) of this equation as a function of p for three different values of β. We can see that it is a well-behaved function of p and thus, there exists a unique value of p that makes it equal to zero. Furthermore, we also observe that all the curves cross the zero line at the same point which gives us a hint that the solution to the equation above is independent of β. The solution is given by the following equation:

$$p = f \left\{ 1 - \left(\frac{1}{f-1} \right)^{\frac{1}{f}} \right\}$$

(3.43)

Next, we derive this equation. Applying the first-order Taylor series expansion on $\mu\{q_1\}$, we get $\mu\{q_1\} = E[Y_1]/E[R_1]$. Using the expressions of $E[Y_1]$ and $E[R_1]$ from (3.2) and (3.4) respectively, we can express $\mu\{q_1\}$, $\mu\{q_1^+\}$, and $\mu\{q_1^-\}$ as follows:

$$\mu\{q_1\} = \frac{fq_1}{q_1\left(q_1 + f(1-q_1)\right)}$$

(3.44)

$$\mu\{q_1^+\} = \frac{fq_1^+}{q_1^+\left(q_1^+ + f(1-q_1^+)\right)}$$

(3.45)

$$\mu\{q_1^-\} = \frac{fq_1^-}{q_1^-\left(q_1^- + f(1-q_1^-)\right)}$$

(3.46)

Substituting these expressions in (3.42), we get:

$$\frac{2}{q_1 + f(1-q_1)} - \frac{1}{q_1^+ + f(1-q_1^+)} - \frac{1}{q_1^- + f(1-q_1^-)} = 0$$

(3.47)

Substituting the value of q_1, q_1^+, and q_1^- from (3.1), (3.32), and (3.33) respectively, into the equation above, and to simplify the presentation, using $\eta = (1 - \frac{p}{f})^t$, we get:

$$\frac{2}{1 - \eta + f\eta} - \frac{1}{1 - \eta^{1+\beta} + f\eta^{1+\beta}} - \frac{1}{1 - \eta^{1-\beta} + f\eta^{1-\beta}} = 0$$

(3.48)

Next, we do algebraic simplification of the expression above.

$$-(1 - \eta + f\eta)\left\{ 1 - \eta^{1+\beta} + f\eta^{1+\beta} + 1 - \eta^{1-\beta} + f\eta^{1-\beta} \right\}$$
$$+ 2\left(1 - \eta^{1+\beta} + f\eta^{1+\beta}\right)\left(1 - \eta^{1-\beta} + f\eta^{1-\beta}\right) = 0$$

(3.49)

Dividing the equation above by $\eta^{1-\beta}$, we get:

$$-(1 - \eta + f\eta)\left\{ 2\eta^{\beta-1} - \eta^{2\beta} + f\eta^{2\beta} - 1 + f \right\}$$
$$+ 2\left(1 - \eta^{1+\beta} + f\eta^{1+\beta}\right)\left(\eta^{\beta-1} - 1 + f\right) = 0$$

(3.50)

Simplifying the equation above, we get:

$$(f - 1) + \eta^{2\beta}\left(-1 + f + 2f\eta - \eta - f^2\eta\right)$$
$$+ 2\eta^{\beta}\left(\eta(f-1)^2 + 1 - f\right) - \eta\left(1 - 2f + f^2\right) = 0$$
$$\Rightarrow (f - 1) + \eta^{2\beta}\left((f-1) - \eta(f-1)^2\right)$$
$$+ 2\eta^{\beta}\left(\eta(f-1)^2 - (f-1)\right) - \eta(f-1)^2 = 0$$

(3.51)

Dividing the equation above by $f - 1$ and simplifying, we get:

$$\Rightarrow (1 - \eta(f - 1))(1 - \eta^\beta)^2 = 0 \tag{3.52}$$

In the equation above, either $1 - \eta(f - 1) = 0$ and/or $1 - \eta^\beta = 0$. The value of $1 - \eta^\beta$ equals zero only when $\beta = 0$, but we know from our problem statement that $\beta \in (0, 1]$ i.e., $\beta \neq 0$. Therefore, $1 - \eta(f - 1) = 0$. Putting back $\eta = (1 - \frac{p}{f})^t$ and solving $1 - (f - 1)(1 - \frac{p}{f})^t = 0$ for p, we get (3.43). Note that this equation does not involve β, which shows that indeed the solution to (3.42) is independent of β as we had intuitively inferred from Figure 3.4.

Equation (3.43) is first of the three equations that we will solve simultaneously. This equation requires the value of actual tag population size t which we do not know. Fortunately, we can calculate an upper bound, t_m, on the actual tag population size and use that in (3.43) instead of t. We will describe a method to obtain t_m in Section 3.5.4, and also determine how close t_m has to be to t to ensure that ART achieves the required reliability.

3.5.2 Number of rounds n

Using the persistence probability calculated in (3.43), the two equations in (3.41) hold. From them, we get:

$$\left(\frac{k\sigma\{q_1\}}{\mu\{q_1^+\} - \mu\{q_1\}} \right)^2 = n = \left(\frac{-k\sigma\{q_1\}}{\mu\{q_1^-\} - \mu\{q_1\}} \right)^2 \tag{3.53}$$

Let Φ be the cumulative distribution function of a standard normal distribution and erf$\{.\}$ be the standard error function, we get:

$$P\{-k \leq Z \leq k\} = \Phi(k) - \Phi(-k) = \text{erf}\left\{ \frac{k}{\sqrt{2}} \right\} \tag{3.54}$$

$P\{-k \leq Z \leq k\}$ gives the success probability in terms of the area under the standard normal curve between $-k$ and $+k$. As per the second of the three conditions, this area should be at least equal to the required reliability α i.e.:

$$P\{-k \leq Z \leq k\} = \alpha \tag{3.55}$$

From (3.54) and (3.55), we get:

$$k = \sqrt{2}\,\text{erf}^{-1}\{\alpha\} \tag{3.56}$$

From (3.53) and (3.56), we get:

$$\left(\frac{\sqrt{2}\,\text{erf}^{-1}\{\alpha\} \times \sigma\{q_1\}}{\mu\{q_1^+\} - \mu\{q_1\}} \right)^2 = n = \left(\frac{-\sqrt{2}\,\text{erf}^{-1}\{\alpha\} \times \sigma\{q_1\}}{\mu\{q_1^-\} - \mu\{q_1\}} \right)^2 \tag{3.57}$$

Equation (3.57) is second of the three equations that we will solve simultaneously.

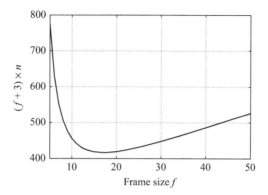

Figure 3.5   Total estimation time vs. frame size

3.5.3   Optimal frame size f

As per the third of the three conditions, total estimation time should be minimum. The total estimation time is directly proportional to total number of slots, $(f + l) \times n$, which is a convex function of f as seen from Figure 3.5. This means that an optimal frame size f_{op} exists and can be obtained by differentiating $(f + l) \times n$ with respect to f as shown below:

$$\frac{d}{df}\{(f + l) \times n\} = 0 \tag{3.58}$$

Equation (3.58) is third of the three equations that we will solve simultaneously.

Required reliability α and confidence interval β are given constants and t_m is calculated using method proposed in Section 3.5.4. Thus, p, q_1, q_1^+, and q_1^- are all functions of f. Consequently, n is a function of f and, therefore, $(f + l) \times n$ is also a function of f with only one unknown, i.e., f. The numerical solution of (3.58) gives the optimal value of frame size, represented by f_{op}.

To numerically solve (3.58), we substitute the value of n from (3.57) in (3.58). As both expressions for n given in (3.57) have same values when p is calculated using (3.43), either of them can be used to calculate n. Substituting n in (3.58) by the L.H.S of the expression for n in (3.57), we get:

$$c[\mu\{q_1^+\} - \mu\{q_1\}]\left[\sigma\{q_1\} + 2(f + l)\frac{\partial\sigma\{q_1\}}{\partial f}\right]$$

$$-2(f + l)\sigma\{q_1\}\left[\frac{\partial\mu\{q_1^+\}}{\partial f} - \frac{\partial\mu\{q_1\}}{\partial f}\right] = 0 \tag{3.59}$$

where $\frac{\partial\mu\{\cdot\}}{\partial f}$ and $\frac{\partial\sigma\{\cdot\}}{\partial f}$ are obtained through the differentiation of expressions for $E[X_b]$ and $Var(X_b)$ in (3.20) and (3.21), respectively. We solve (3.59) numerically to obtain f_{op}.

Summary of steps to calculate p, n, and f_{op}

First, we calculate the value of t_m, as explained in Section 3.5.4. Second, we numerically solve (3.59) to obtain f_{op}. Third, we put this value of f_{op} along with t_m in (3.43) to obtain the value of p. Last, we put the resulting value of p along with f_{op} in (3.57) and obtain the value of n. Note that although (3.43) does not involve α and β, p still depends on them because it is a function of f and the optimal value of f depends on α and β.

Table 3.2 shows the values of p, n, and f_{op} for different accuracy requirements and tag population sizes calculated using the steps described above. We observe from this table that for a given tag population size, as the value of α increases and/or β decreases, the value of n increases to fulfill the more stringent accuracy requirements. We also observe from this table that for a given (α, β) pair, the values of f_{op} and n are the same for all tag population sizes, which shows that total number of slots, $(f_{op} + 1) \times n$, depends only on the accuracy requirements and is independent of tag population size. We will formally prove the independence of estimation time from tag population size in Section 3.7.1. We further observe that as the tag population size increases, the value of p decreases to reduce the number of tags participating in a frame to keep the value of f_{op} and n independent of tag population size.

3.5.4 Obtaining population upper bound t_m

So far we have assumed the knowledge of an upper bound t_m on tag population size t. We now present a fast scheme to obtain t_m based on Flajolet and Martin's probabilistic counting algorithm [9]. Before calculating system parameters p, n, and f_{op}, the reader uses this scheme to obtain t_m. In this scheme, the reader keeps issuing single-slot frames, where the persistence probability p follows a geometric distribution starting from $p = 1$ (i.e., $p = \frac{1}{2^{i-1}}$ in the ith frame), until the reader gets an empty slot. Suppose the empty slot occurred in the ith frame, then $t_m = 1.2897 \times 2^{i-2}$ is an upper bound on t [9,23]. According to [9], t_m asymptotically approaches t when instead of using a single value of the first empty slot from one experiment, we use average of values of the first empty slot from a large number of experiment.

Next, we determine how close the upper bound t_m has to be to the actual tag population size to ensure that ART achieves the required reliability and examine whether t_m obtained using $t_m = 1.2897 \times 2^{i-2}$ lies close enough to t. We derive an expression to calculate the expected value of actual reliability, denoted by $\tilde{\alpha}$, as a function of t_m given that the required reliability α, confidence interval β, and the actual tag population size t are known.

Equation (3.57) is obtained using the condition that actual reliability should be greater than or equal to the required reliability. Therefore, we use this equation to derive an expression for expected value of actual reliability. In (3.57), we calculate n using q_1, q_1^+, and q_1^-, which are obtained from (3.1), (3.32), and (3.33), respectively, by putting $t = t_m, f = f_{op}$, and $p = f_{op} \left\{ 1 - \left(\frac{1}{f_{op}-1} \right)^{\frac{1}{t_m}} \right\}$. This gives us:

$$q_1 = 1 - \left(\frac{1}{f_{op} - 1} \right), \quad q_1^{\pm} = 1 - \left(\frac{1}{f_{op} - 1} \right)^{1 \pm \beta} \tag{3.60}$$

Table 3.2 Values of f_{op}, n, and p for different values of α, β, and tag population size

Accuracy requirement	Tag population size								
	10^2			10^4			10^6		
	f_{op}	n	p	f_{op}	n	p	f_{op}	n	p
$\alpha = 60.0\%, \beta = 40.0\%$	12	1.00E+00	2.84E−01	12	1.00E+00	2.88E−03	12	1.00E+00	2.88E−05
$\alpha = 70.0\%, \beta = 30.0\%$	14	2.00E+00	3.55E−01	14	2.00E+00	3.59E−03	14	2.00E+00	3.59E−05
$\alpha = 80.0\%, \beta = 20.0\%$	15	4.00E+00	3.91E−01	15	4.00E+00	3.96E−03	15	4.00E+00	3.96E−05
$\alpha = 90.0\%, \beta = 10.0\%$	15	2.50E+01	3.91E−01	15	2.50E+01	3.96E−03	15	2.50E+01	3.96E−05
$\alpha = 95.0\%, \beta = 5.00\%$	15	1.43E+02	3.91E−01	15	1.43E+02	3.96E−03	15	1.43E+02	3.96E−05
$\alpha = 99.0\%, \beta = 1.00\%$	15	6.24E+03	3.91E−01	15	6.24E+03	3.96E−03	15	6.24E+03	3.96E−05
$\alpha = 99.9\%, \beta = 0.10\%$	15	1.02E+06	3.91E−01	15	1.02E+06	3.96E−03	15	1.02E+06	3.96E−05

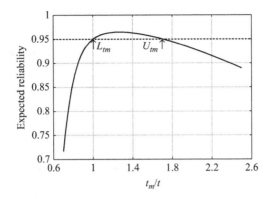

Figure 3.6 Expected value of actual reliability vs. $\frac{t_m}{t}$

As the number of tags in the population are t and not t_m, when the reader executes the frames, the actual values of q_1, q_1^+, and q_1^- represented by $\hat{q}_1$, $\hat{q}_1^+$, and $\hat{q}_1^-$, respectively, follow the equations below.

$$\hat{q}_1 = 1 - \left(\frac{1}{f_{op}-1}\right)^{\frac{t}{t_m}}, \quad \hat{q}_1^\pm = 1 - \left(\frac{1}{f_{op}-1}\right)^{\frac{t}{t_m}(1\pm\beta)} \tag{3.61}$$

Let $\tilde{\alpha}$ represent the expected value of actual reliability in n rounds when the population contains t tags and the calculated value of upper bound is t_m, then the following equality holds.

$$\left(\frac{\sqrt{2}\,\mathrm{erf}^{-1}\{\tilde{\alpha}\} \times \sigma\{\hat{q}_1\}}{\mu\{\hat{q}_1^+\} - \mu\{\hat{q}_1\}}\right)^2 = n = \left(\frac{-\sqrt{2}\,\mathrm{erf}^{-1}\{\tilde{\alpha}\} \times \sigma\{\hat{q}_1\}}{\mu\{\hat{q}_1^-\} - \mu\{\hat{q}_1\}}\right)^2 \tag{3.62}$$

Substituting the value of n from (3.57) into the equation above and solving for $\tilde{\alpha}$, we get:

$$\begin{aligned}
\tilde{\alpha} &= \mathrm{erf}\left\{\mathrm{erf}^{-1}\{\alpha\} \times \frac{\sigma\{q_1\}}{\sigma\{\hat{q}_1\}} \times \frac{\mu\{\hat{q}_1^+\} - \mu\{\hat{q}_1\}}{\mu\{q_1^+\} - \mu\{q_1\}}\right\} \\
&= \mathrm{erf}\left\{\mathrm{erf}^{-1}\{\alpha\} \times \frac{\sigma\{q_1\}}{\sigma\{\hat{q}_1\}} \times \frac{\mu\{\hat{q}_1^-\} - \mu\{\hat{q}_1\}}{\mu\{q_1^-\} - \mu\{q_1\}}\right\}
\end{aligned} \tag{3.63}$$

The expected actual reliability $\tilde{\alpha}$ is a convex function of $\frac{t_m}{t}$ and is equal to α for two values of $\frac{t_m}{t}$ represented by L_{tm} and U_{tm}. Figure 3.6 plots the expected value of actual reliability $\tilde{\alpha}$ as a function of $\frac{t_m}{t}$ using Equation (3.63) with $\alpha = 95\%$ and $\beta = 5\%$. The dashed horizontal line in the figure marks the required reliability $\alpha = 95\%$. The actual reliability will be greater than or equal to the required reliability as long as the value of $\frac{t_m}{t}$ satisfies the following condition:

$$L_{tm} \leq \frac{t_m}{t} \leq U_{tm} \tag{3.64}$$

Table 3.3 U_{tm} for different population sizes and accuracy requirements

Accuracy requirement	Tag population size			
	10^3	10^4	10^5	10^6
$\alpha = 90.00\%, \beta = 10.0\%$	1.83	1.83	1.83	1.83
$\alpha = 95.00\%, \beta = 5.00\%$	1.71	1.71	1.71	1.71
$\alpha = 99.00\%, \beta = 1.00\%$	1.66	1.66	1.66	1.66
$\alpha = 99.90\%, \beta = 0.10\%$	1.64	1.64	1.64	1.64
$\alpha = 99.99\%, \beta = 0.01\%$	1.64	1.64	1.64	1.64

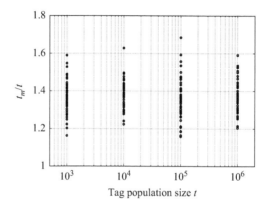

Figure 3.7   Experimentally observed values of ratio $\frac{t_m}{t}$

The values of L_{tm} and U_{tm} can be obtained by using $\tilde{\alpha} = \alpha$ in (3.63) and solving it for t_m and dividing it by the tag population size t. This results in two values of $\frac{t_m}{t}$ because $\tilde{\alpha}$ is a convex function of $\frac{t_m}{t}$ and its maxima is greater than α. The value of L_{tm} is always equal to 1 and the value of U_{tm} is calculated by the numerical solution of (3.63) using $\tilde{\alpha} = \alpha$.

The value of U_{tm} depends on the required reliability α and confidence interval β. Table 3.3 tabulates the values of U_{tm} for different population sizes and accuracy requirements. We observe from Table 3.3 that the value of U_{tm} is independent of tag population size. This is because U_{tm} depends on q_1 for a given α and β (according to (3.63)) and q_1 is independent of tag population size as we will discuss in Section 3.7.1. We also observe that U_{tm} decreases with increasing accuracy requirements. This makes intuitive sense because the higher the required accuracy, the lesser the error in the upper bound t_m that can be tolerated. We see from Table 3.3 that even for very high accuracy requirements of $\alpha = 99.99\%$ and $\beta = 0.01\%$, the value of t_m calculated as $t_m = 1.2897 \times 2^{i-2}$ can be up to $1.64 \times t$.

From simulations, we have observed that the value of t_m calculated as $t_m = 1.2897 \times 2^{i-2}$ always lies within t and $1.64 \times t$. This is seen in Figure 3.7, where we plot the observed values of $\frac{t_m}{t}$ obtained through 100 runs of simulations using

$t_m = 1.2897 \times 2^{i-2}$ for different values of tag population size. Within each simulation run, we obtained 10 values of i, averaged them, and replaced i with that average in the equation $t_m = 1.2897 \times 2^{i-2}$ to obtain $\frac{t_m}{t}$.

3.6 ART—practical considerations

In this section, we describe how ART estimates sizes of arbitrarily large tag populations. We also present the method that ART employs to enable the use of multiple RFID readers for estimating the size of a given RFID tag population.

3.6.1 Unbounded tag population size

For a given value of frame size f, Theorem 3.2 calculates the upper bound t_M on the number of tags that ART can estimate. This upper bound exists because for tag population sizes larger than t_M, the system parameters take on values that cannot be implemented practically. After Theorem 3.2, we describe how we extend ART to estimate sizes of arbitrarily large populations.

Theorem 3.2. *For a given frame size $f > 1$, the maximum number of tags t_M that ART can estimate is*

$$t_M = -\frac{\ln\{f - 1\}}{\ln\{1 - \frac{1}{2^{15}}\}} \tag{3.65}$$

Proof. In theory, we can increase the estimation scope of ART to any population size by decreasing the value of p according to (3.43). In practice, however, f/p has a minimum value of $2^{15} - 1$. Recall that in ART, the reader announces a virtual frame size of f/p (although terminates the frame after the first f slots) and each tag uses the result of a hash function h to select a slot in the range $[1, f/p]$. The number of bits to store the result of the hash function is specified to be 15 in the C1G2 standard. Thus, the maximum value of f/p can be $2^{15} - 1$, i.e.:

$$p > \frac{f}{2^{15}} \tag{3.66}$$

Substituting the value of p from (3.43) into the equation above, we get:

$$f\left\{1 - \left(\frac{1}{f-1}\right)^{\frac{1}{t}}\right\} > \frac{f}{2^{15}} \tag{3.67}$$

Rearranging the expression above and solving for t, we get:

$$t < -\frac{\ln\{f - 1\}}{\ln\{1 - \frac{1}{2^{15}}\}} = t_M \tag{3.68}$$

$\square$

As an example, with $f = 15$, t_M is just 86,475. Practically, ART achieves required reliability only for tag populations smaller than t_M. If population size is larger than t_M, ART requires $p \leq \frac{f}{2^{15}}$, which is practically not possible with C1G2 RFID tags. This limitation exists with all the existing estimation schemes but has never been addressed before.

Next, we present a strategy to estimate the sizes of arbitrarily large tag populations. The key idea is to first divide the entire population into smaller subpopulations of roughly equal sizes and then estimate the size of each subpopulation independently. At the end, adding the estimated sizes of all subpopulations gives the estimate of number of tags in the entire population. The size of any subpopulation should not require $\frac{f}{p} \geq 2^{15}$.

Next, we first calculate the number of subpopulations that ART should divide a given tag population into and then present a strategy to perform this division virtually (i.e., requiring no manual division of tags). Maximum number of tags that a subpopulation can have is given by (3.65). Therefore, the minimum number of subpopulations that the entire tag population should be divided into is $\frac{t_m}{t_M}$, where t_m is calculated as explained in Section 3.5.4.

To divide the tag population into subpopulations, we use the SELECT command standardized in the C1G2 standard. The ID of a tag is stored in its memory at a specific memory address. The tag can retrieve any bits stored in its memory by specifying an appropriate address range. Using the SELECT command, a reader can broadcast an address range and a bit mask that specifies which tags should participate in an Aloha frame. Each tag compares the bit mask with the bits in the specified address range in its memory and participates in the frame only if the bit mask matches the specified bits in its memory. To divide the whole population into subpopulations of roughly equal sizes, we leverage the fact that in large populations, the expected number of tags whose IDs have the least significant bit (LSB) of 0 is approximately the same as the expected number of tags whose IDs have the LSB of 1. Similarly, the expected number of tags whose IDs have the two LSBs of 00 is approximately the same as the expected number of tags whose IDs have the two LSBs of 01, 10, or 11, and so on. Therefore, a reader can divide the tag population into 2^z groups of roughly equal sizes by specifying appropriate masks for the address range corresponding to the z LSBs of tag IDs. The value of z is given by $\left\lceil \log_2 \left\{ \frac{t_m}{t_M} \right\} \right\rceil$.

To summarize, a reader first obtains the value of upper bound t_m. Second, it calculates the value of n and f_{op}. Third, it calculates the value of t_M using (3.65). Fourth, it calculates $z = \left\lceil \log_2 \left\{ \frac{t_m}{t_{max}} \right\} \right\rceil$. Fifth, it executes 2^z independent estimation rounds for required reliability α and confidence interval β, where in each round it uses SELECT command with a unique z bit mask for the z LSBs of the tag IDs. In each independent estimation round, it uses $p = f_{op} \left\{ 1 - \left(\frac{1}{f_{op}-1} \right)^{\frac{1}{t_m/2^z}} \right\}$. Finally, it adds up all 2^z estimates to obtain the estimate of total number of tags in the population.

3.6.2 ART with multiple readers

We next discuss how to obtain t_m and $\tilde{t}$ using multiple readers with overlapping coverage. To obtain t_m using multiple readers, we can let each reader obtain the t_m value on its own and then sum them up as the final overall t_m because of two reasons. First, our requirement on t_m is only a rough upper bound with an error tolerance of over $1.64 \times t$. Second, deployment of multiple readers in practice often requires site surveys to ensure minimal overlapping between readers.

To obtain $\tilde{t}$ using multiple readers, we adapt the approach proposed by Kodialam *et al.* in [54], which uses a central controller for all readers. ART parameters β, α, t_m, p, n, and f_{op} have the same value across all readers. When a reader transmits seed R_i in its ith frame, it does not generate R_i on its own, rather it uses the ith seed R_i issued by the central controller. That is, each reader generates the same sequence of n seeds. In the ith frames from different readers, because all readers use the same seed R_i, the slot number that a given tag chooses is the same (i.e., $h(f, R_i, ID)$) in the frame of each reader covering this tag. Once a reader has completed its frame, it sends the frame to the central controller. The controller applies the logical OR on all the ith frames from all readers, and gets a single ith frame as if using a single reader. ART uses the n frames computed by logical OR to estimate the population size. Pseudocode of ART is given in Algorithms 1 and 2.

3.7 ART—analysis

In this section, first, we prove that the estimation time of ART is independent of the tag population size. Second, we briefly discuss the computational complexity of ART. Last, we perform an analytical comparison of ART with existing schemes to mathematically justify the faster speed of ART compared to existing schemes.

3.7.1 Independence of estimation time from tag population size

There are three inputs to ART: confidence interval β, required reliability α, and a population of t tags where t is unknown. The total number of slots of ART, $(f_{op} + l) \times n$, actually does not depend on t. Intuitively, the larger t is, the smaller p is according to (3.43). Although t plays an important role in computing p, n, and f individually, in formula $(f_{op} + l) \times n$ the impact of t eventually gets canceled out. Next, we prove this independence.

From (3.57), we observe that the value of n depends on α, β, μ, σ and from (3.59), we observe that the value of f_{op} depends upon β, μ, σ. Thus, the total number of slots $(f_{op} + l) \times n$ depends on α, β, μ, σ. The values of α and β are given constants and μ and σ are functions of q_1, as seen from (3.20) and (3.21). To prove that $(f_{op} + l) \times n$ is independent of t, we have to prove that q_1 is independent of t. From (3.1), we have $q_1 = 1 - (1 - \frac{p}{f})^t$. As we do not know the value of t, rather we know t_m, we use $q_1 = 1 - (1 - \frac{p}{f})^{t_m}$. Substituting the value of p using $t = t_m$ from (3.43) into this expression of q_1, we get:

Algorithm 1: Estimate RFID Tag Population (α, β, n_r)

    Input: (1) Required reliability α
            (2) Required confidence interval β
            (3) Number of readers n_r
    Output: Estimated tag population size $\tilde{t}$

1:   $t_m := $ **CalculateUpperBound**(n_r)
2:   Solve (3.59) to get f_{op}.
3:   Put the value of f_{op} and t_m in (3.43) to get p.
4:   Evaluate n by using α, β, f_{op}, and p, in (3.57).
5:   Evaluate t_M using (3.65).
6:   **if** $t_m > t_M$ **then**
7:       $z := \left\lceil \log_2\left\{\frac{t_m}{t_M}\right\}\right\rceil$
8:   **end**
9:   **else**
10:      $z := 0$
11:   **end**
12:   Reevaluate p by using $p = f_{op}\left\{1 - \left(\frac{1}{f_{op}-1}\right)^{\frac{1}{t_m/2^z}}\right\}$.
13:   $\tilde{t} := 0$
14:   **for** $j := 1$ *to* 2^z **do**
15:       Use SELECT with z bit binary representation of j as mask.
16:       **for** $i := 1$ *to* n **do**
17:            Provide all readers with f_{op}/p and random seed R_i.
18:            Run Aloha on readers and gather all readers' frames.
19:            Perform slot wise OR on all frames to get one frame.
20:            Obtain $\tilde{X}_1(i)$, the average run size of 1s in this frame.
21:       **end**
22:       $\tilde{X}_{avg} \leftarrow \sum_{i=1}^{n} \tilde{X}_1(i)/n$
23:       Use $E[X_1] := \tilde{X}_{avg}$ and solve (3.20) to obtain an estimate $\tilde{t}_z$ of the number of tags in the current subpopulation.
24:       $\tilde{t} := \tilde{t} + \tilde{t}_z$
25:   **end**
26:   **return** $\tilde{t}$

$$q_1 = 1 - \left(1 - \frac{1}{f} \times f\left\{1 - \left(\frac{1}{f-1}\right)^{\frac{1}{t_m}}\right\}\right)^{t_m} = \frac{f-2}{f-1} \tag{3.69}$$

Thus, the value of q_1 that we use to calculate μ and σ and consequently f_{op} and n is independent of tag population size t or the upper bound on tag population size t_m. Therefore, f_{op} and n depend only on α and β regardless of the value of t or t_m. The upper bound on tag population size t_m only affects the value of p. For ART to achieve

---

Algorithm 2: Calculate Upper Bound (n_r)

    Input: Number of readers n_r
    Output: Upper bound on tag population size t_m
1: $f := 1$
2: **for** $j := 1$ *to* n_r **do**
3:     $i := 1$
4:     $p_i := 1$
5:     **repeat**
6:         Provide reader j with f/p_i and a random seed R_i.
7:         Run Aloha on reader j and get the response.
8:         **if** *slot is not empty* **then**
9:             $p_{i+1} := p_i/2$
10:             $i := i + 1$
11:         **end**
12:     **until** *slot is empty*
13:     $t_{m,j} := 1.2897 \times 2^{i-2}$
14: **end**
15: $t_m := \sum_{j=1}^{n_r} t_{m,j}$
16: **return** t_m

---

the required reliability, this upper bound has to satisfy the condition $L_{tm} \le t_m \le U_{tm}$. If $t_m > t \times U_{tm}$, the required reliability will not be achieved because the value of p will become so small that enough number of tags will not participate in the frames. Regardless, the value of $(f_{op} + 1) \times n$ stays the same. We have seen from Figure 3.7 that for all practical purposes, the value of t_m satisfies the requirement $L_{tm} \le t_m \le U_{tm}$ when calculated using the method proposed in Section 3.5.4.

3.7.2 Computational complexity

The two most computationally intensive tasks in ART are the numerical solutions of (3.20) to obtain the estimate $\tilde{t}$ and of (3.59) to calculate f_{op}. Fortunately, these two equations need to be solved numerically only once during the estimation process: (3.59) before executing the frames and (3.20) after executing the frames. Consequently, the runtime complexity of ART is no larger than that of a standardized Aloha protocol. Almost all existing schemes involve numerical solutions of equations to obtain the estimate $\tilde{t}$. Therefore, the off-line computational complexity of ART is comparable to those of existing estimation schemes.

3.7.3 Analytical comparison of estimators

Next, we show that the ART estimator, namely the average run size of 1s, has less variance than many other framed-slotted Aloha based estimators, namely (1) the size of the first run of 0s (used by FNEB [11]), (2) the average run size of 0s, (3) the total number of 0s (used by UPE [50] and EZB [54]), (4) the total number of 1s, (5) the total

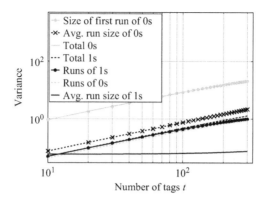

Figure 3.8   Variance of different estimators vs. RFID tag population size

number of runs of 0s, and (6) the total number of runs of 1s. The higher the variance of an estimator is, the more number of rounds n are needed to improve reliability, and more rounds means larger estimation time. Figure 3.8 shows the analytical plots of the variances of the ART estimator and the above six estimators with frame size $f = 16$ versus tag population sizes. This figure shows that *the variance of ART estimator is significantly lower than all other estimators*. Runs of 1s and runs of 0s have smaller variance compared to ART for very small tag population sizes. This observation, however, is insignificant because both these quantities are nonmonotonic functions of tag population size and therefore, cannot be used alone for estimation. The variances of these estimators are calculated as follows. The variance of the total number of 0s and 1s is calculated using (3.3). The variance of the size of the first run is calculated using Equation (3) in [45] by setting $i = 1$. The variance of the number of runs of 0s and that of 1s is calculated using (3.5). We emphasize that plots in Figure 3.8 are not based on experimental results, instead, they are based on analytical formulas.

3.8   Performance evaluation

We numerically evaluated in MATLAB® our ART scheme as well as four prior RFID estimation schemes: UPE [50], EZB [54], FNEB [11], and MLE [75]. We did not evaluate LoF [23] because it is noncompliant with C1G2 and CSE [58] because it does not take accuracy requirements as input. The estimation times for ART reported in this section include the time required to obtain the value of t_m. To ensure compliance with the C1G2 standard, in all our simulations, each tag picks up exactly one slot at the start of frame as soon as the reader broadcasts the frame size.

Next, we first conduct a side-by-side comparison on estimation time between ART and the four prior schemes. Then, we conduct experiments to show that ART indeed achieves the required reliability.

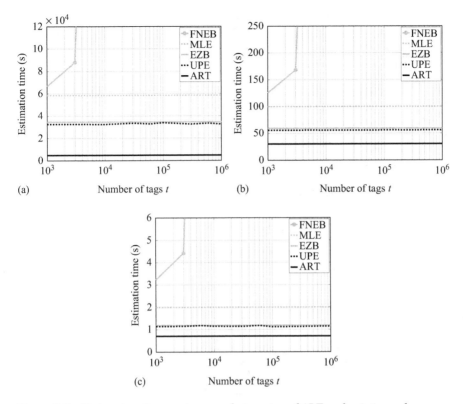

Figure 3.9 Estimation time vs. tag population size of ART and existing schemes
for three different accuracy requirements: (a) $\alpha = 95\%$, $\beta = 5\%$,
(b) $\alpha = 99\%$, $\beta = 1\%$, and (c) $\alpha = 95\%$, $\beta = 5\%$

3.8.1 Estimation time

The results in Figures 3.9–3.11 show that *the estimation time of ART is significantly
smaller than all prior schemes*. Note that in Figures 3.10 and 3.11, the plots for FNEB
are out of the range of the vertical axes, and the plots of UPE and EZB are almost
overlapping.

We make three main observations from Figure 3.9(a)–(c), which show the estimation time needed by each scheme with population sizes of up to one million tags for
different configurations of α and confidence interval β. First, we observe that ART
is faster than all four prior schemes in all these configurations. For $\alpha = 99.9\%$ and
$\beta = 0.1\%$, ART is seven times faster than the fastest prior estimation schemes, which
are UPE [50] and EZB [54]. For $\alpha = 99\%$ and $\beta = 1\%$, ART is 1.96 times faster
than UPE and EZB. For $\alpha = 95\%$ and $\beta = 5\%$, ART is 1.68 times faster than UPE
and EZB. Second, we observe that ART, UPE, EZB, and MLE perform estimation
in constant time, which attributes to the use of persistence probabilities. Third, we
observe that FNEB, whose estimator is the size of the first run of 0s, is the slowest.

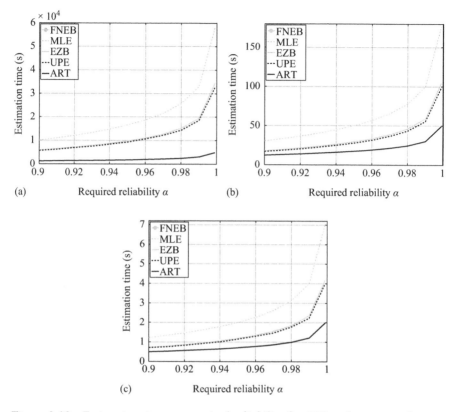

*Figure 3.10   Estimation time vs. required reliability for ART and existing schemes
for three different confidence intervals: (a) β = 0.1%, (b) β = 1%,
and (c) β = 5%*

This concurs with our analytical analysis in Figure 3.8, where we show that FNEB
has the largest variance. The larger the variance of an estimator, the more the rounds
of execution needed to achieve the required reliability, and the longer the estimation
time.

We make three main observations from Figure 3.10(a)–(c), which show the estimation time of each scheme for 5, 000 tags with the required reliability α varying from
90% to 99.9% for different configurations of confidence interval β. First, we observe
that ART is faster than all four prior estimation schemes in all these configurations.
Second, the difference between the estimation time of ART and those of prior schemes
increases as the required reliability increases. For example, for $\beta = 5\%$ and $\alpha = 95\%$,
ART is 1.68 times faster than UPE and EZB, while for $\beta = 0.1\%$ and $\alpha = 99.9\%$,
it is 7 times faster. This shows that ART becomes more and more advantageous over
existing schemes when the required reliability increases. Third, for all schemes, the
estimation time increases as the required reliability increases because more number
of rounds are needed to achieve the required reliability. We further observe that ART's

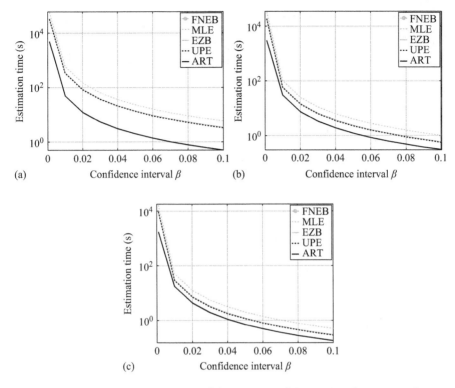

Figure 3.11   *Estimation time vs. confidence interval for ART and existing schemes*
for three different required reliability values: (a) $\alpha = 99.9\%$,
(b) $\alpha = 99\%$, and (c) $\alpha = 95\%$

estimation time increases at the lowest rate as the required reliability increases because
its estimator has the smallest variance.

We make three main observations from Figure 3.11(a)–(c), which show the estimation time of each scheme for 5,000 tags with the confidence interval β varying from
0.1% to 10% for different configurations of α. First, we observe that ART is faster
than all estimation schemes in all these configurations. Second, for all schemes, the
estimation time decreases as the confidence interval increases because lesser number
of rounds are needed to achieve the required reliability.

3.8.2   Actual reliability

The part figures in Figure 3.12 show the actual reliability of ART versus the number of
tags for different configurations of required reliability α and confidence interval β.
We observe that *ART always achieves the required reliability*. These figures show
several ups and downs in the plotted values. These ups and downs are not because of
any noise, rather we see them because of the magnification level of vertical axis in
these figures.

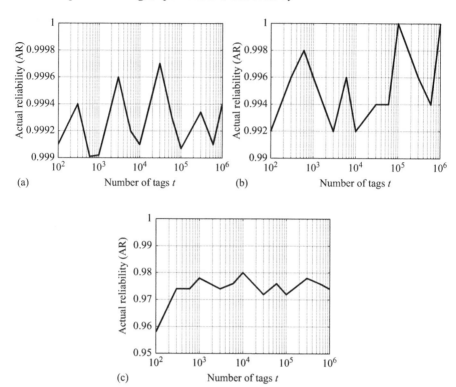

Figure 3.12   Actual reliability achieved by ART for three different required
reliability values and confidence intervals: (a) α = 99.9%, β = 0.1%,
(b) α = 99%, β = 1%, and (c) α = 95%, β = 5%

3.9   Conclusion

The key technical novelty of this chapter is in proposing the new estimator, the average
run size of 1s, for estimating RFID tag population size of arbitrarily large sizes. Using
analytical plots, we show that our estimator has much smaller variance compared to
other estimators including those used in prior work. It is this smaller variance that
makes our scheme faster than the previous ones. The key technical depth of this chapter
is in the mathematical development of the estimation theory using this estimator. ART
can estimate arbitrarily large tag populations with arbitrarily high accuracy. It works
with single as well as multiple readers. Our experimental results show that ART is
significantly faster than all prior RFID estimation schemes. We have shown, both
theoretically and experimentally, that the estimation time of ART is independent of
the tag population size.

Chapter 4

RFID estimation—impact of blocker tags

4.1 Introduction

4.1.1 Background and motivation

Radio-frequency identification (RFID) technique has risen to be a revolutionary element in supply chain management and inventory control [60–68], as the cost of commercial passive RFID tags is negligible compared with the value of the products to which they are attached (e.g., as low as 5 cents per tag [24]). For example, in Hong Kong International Airport where RFID systems are used to track shipment, the average daily cargo tonnage in May 2010 was 12K tonnes and has been on the rise [69]. As real-time information is made available, the administration and planning processes can be significantly improved. An RFID system typically consists of a reader and a population of tags [70]. A reader has a dedicated power source with significant computing capability. It transmits commands to query a set of tags, and the tags respond over a shared wireless medium. A tag is a microchip with an antenna in a compact package that has limited computing capability and a longer communication range than barcodes. There are two types of tags: *passive* that do not have their own power sources and are powered up by harvesting the radio frequency energy from readers and *active* that have their own power sources.

An inevitable fact is that the widely used RFID tags impose serious privacy concerns, as when a tag is interrogated by an RFID reader; no matter whether the reader is authorized or not, it blindly responds with its ID and other stored information (such as manufacturer, product type, and price) in a broadcast fashion. For example, a woman may not want her dress sizes and a patient may not want his/her medication, to be publicly known. Some cryptography-based authentication protocols have been proposed to circumvent malicious scanning [71]. However, none of these protocols is compliant with the C1G2 standard. Furthermore, these protocols often require computational resources that exceed the capability of commercial C1G2-compliant passive RFID tags. If such additional computational capability is indeed implemented, the cost of such tags will be much more expensive than the C1G2-compliant passive tags. Hence, we use blocker tags, which are easy to deploy, to protect RFID privacy. A *blocker tag* is an RFID device that is preconfigured with a set of known RFID tag IDs, which we call *blocking IDs*. The blocker tag behaves as if all tags with its blocking IDs are present. A blocker tag protects the privacy of the set of genuine tags

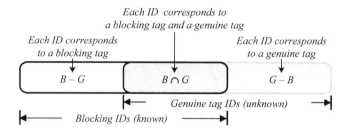

Figure 4.1   Three types of IDs in the system containing blocker tags

whose IDs are among the blocking IDs of the blocker tag because any response from a genuine tag is coupled with the simultaneous response from the blocker tag; thus, the two responses collide, and attackers cannot obtain private information.

4.1.2   Problem statement

This chapter concerns with the problem of RFID estimation with the presence of a blocker tag. Formally, the problem is defined as follows. *Given (1) a set of unknown genuine tags G of unknown size g, (2) a blocker tag with a set of blocking IDs B, which is configured by the system manager, (3) a required confidence interval $\alpha \in (0, 1]$, and (4) a required reliability $\beta \in [0, 1)$, we want to estimate the number of genuine tags in G, denoted as $\hat{g}$, so that $P\{|\hat{g} - g| \leq \alpha g\} \geq \beta$. Besides time efficiency, we also take the energy efficiency into consideration if the battery-powered active RFID tags are used.* We assume that the blocker tag is trusted because the manager is in control of the blocker tag. Hence, the blocking ID set B is known by the manager in prior. In contrary, for the genuine tag set G, we know neither its size nor the exact tag IDs in it. As shown in Figure 4.1, the sets B and G may overlap.

This problem may arise in many applications. Consider an RFID-enabled logistics center, where each package is affixed with an RFID tag that contains the delivery address and the item information. The information of some packages, e.g., the medicine someone purchased on Amazon, are closely related to the customers' privacy. To protect their personal privacy, we use a blocker tag to prevent the tags from malicious scanning. Meanwhile, the manager may want to use an RFID estimation protocol to monitor the number of packages for the purpose of making an efficient delivery plan. How about turning off the blocker tag and then using prior RFID estimation schemes to estimate the number of genuine tags? Turning off the blocker tag will give attackers a time window to breach privacy, especially for the scenarios in which RFID estimation schemes are being continuously performed for monitoring purposes.

4.1.3   Limitations of prior art

To the best of our knowledge, this chapter is the first to investigate RFID estimation with the presence of a blocker tag. Although some RFID estimation schemes have

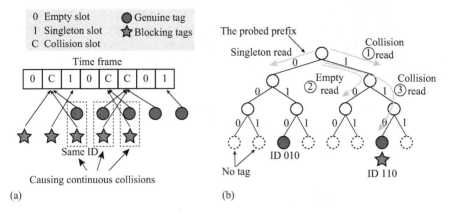

Figure 4.2   Exemplify the impact of blocker tag on the tag identification protocols.
(a) Aloha-based protocols. (b) Tree-based protocols

been proposed [50,54,70,72–76], none of them considers the presence of a blocker tag. Furthermore, none of them can be easily adapted to solve our problem. For example, the state-of-the-art average run-based tag (ART) estimation protocol uses the framed-slotted Aloha communication mechanism specified in the C1G2 standard. The reader queries the tags by initializing a slotted frame, and each tag randomly selects a slot to reply the response. ART leverages the average run length of nonempty slots observed from the time frame to estimate the tag cardinality. Clearly, ART can only tell the tag cardinality of the universal set $U = B \cup G$, which, however, is not what we want. Due to the same reason, all the other existing tag estimation protocols cannot address the new problem of RFID estimation with the presence of a blocker tag.

How about using the tag identification protocols? Generally, there are two categories of identification protocols: Aloha-based protocols [77] and tree-based protocols [78]. It is a well-recognized fact that the identification protocols are slow because their execution time is proportional to the tag population. What is worse, their efficiency will further deteriorate with the presence of a blocker tag. Their basic principles can be found in Section 4.5. Here, we only elaborate why these two types of identification protocols become more inefficient with the presence of a blocker tag. As exemplified in Figure 4.2(a), the responses from two tags with the same ID in $B \cap G$ always collide with each other. Thus, the genuine tag IDs in $B \cap G$ can never be identified. Moreover, the large number of *continuous collisions* also seriously hinders the identification of the tags in $(B - G) \cup (G - B)$.

The tree-based identification protocols can identify the IDs in $(B - G) \cup (G - B)$ when a queried prefix is followed by a successful read; and identify the IDs in $B \cap G$ when a prefix whose length is equal to tag ID but still followed by a collision read. Then, we can get the set G, by calculating $\{(B - G) \cup (G - B) - B\} \cup (B \cap G)$. The cardinality g is obtained upon getting G. When there is blocker tag in the system, the tree-based identification becomes slow because the continuous collisions caused by the tags in $B \cap G$ always "lure" the reader to continuously extend the probed prefix by 1 bit each time, until the probed prefix reaches the ID length. As exemplified in Figure 4.2(b), to identify the ID "010," the reader only needs to probe one prefix "0"; however, to identify the ID "110" (an ID in $B \cap G$), the reader needs to successively probe three prefixes: "1"→ "10"→"110" (until the prefix length is equal to ID length). Since the length of prefix "110" is already equal to the length of tag ID, we know the ID "110" is necessary to correspond to two tags: one of them is a blocking tag, and the other one is a genuine tag. We observed from the simulation results that the identification protocols are seriously slow, and their time cost is even hundreds of times longer than REB.

4.1.4   Proposed approach

In this chapter, we propose a *RFID estimation scheme with blocker tags* (*REB*). The communication protocol used by REB is the standard framed-slotted Aloha protocol, in which a reader first broadcasts a value f and a random number R to the tags, where f represents the number of time slots in the forthcoming frame. Then, each tag computes a hash using the random number R and its ID, where the resulting hash value h is within $[0, f - 1]$, and the tag replies during slot h. For each slot, if the reader does not receive any tag response, we represent this slot as 0; if the reader successfully receives a tag response, we represent this slot as 1; if the reader senses the collided tag responses (two or more tags respond simultaneously), and we represent this slot as c. Note that a reader can detect if there is a collision according to the C1G2 standard. Executing this protocol for the blocking IDs (simulated by the blocker tag) and genuine tags, we get a ternary array $\mathbb{BG}[0 \ldots f - 1]$ where each bit is 0, 1, or c. As we know the blocking IDs, we can virtually execute the framed-slotted Aloha protocol using the same frame size f and random number R for the blocking IDs; thus, we get a ternary array $\mathbb{B}[0 \ldots f - 1]$ where each bit is 0, 1, or c. Here, if no blocking ID is hashed to this position, it is represented by 0; if only one blocking ID is hashed to this position, it is represented by 1; if two or more blocking IDs are hashed to this position, it is represented by c. From the two arrays $\mathbb{BG}[0 \ldots f - 1]$ and $\mathbb{B}[0 \ldots f - 1]$, REB counts two numbers: N_{00}, which is the number of slots i such that both $\mathbb{BG}[i] = 0$ and $\mathbb{B}[i] = 0$, and N_{11}, which is the number of slots i such that both $\mathbb{BG}[i] = 1$ and $\mathbb{B}[i] = 1$. REB is based on the key insight that in general the smaller N_{00} is, the larger $|B \cup G|$ is and the larger N_{11} is, the larger $|B - G|$ is. In this chapter, we establish a monotonous functional relationship between N_{00} and $|B \cup G|$, and a monotonous functional relationship between N_{11} and $|B - G|$. Thus, from the observed N_{00} and N_{11}, we can estimate $|B \cup G|$ and $|B - G|$. Then, we can calculate the size of G because $|G| = |B \cup G| - |B - G|$.

4.1.5   Challenges and proposed solutions

The first challenge is to guarantee the required estimation accuracy that is specified by the confidence interval $\alpha \in (0, 1]$ and the reliability $\beta \in [0, 1)$. The estimator is not precise due to its probabilistic nature. Since a single frame is usually not able to output an accurate estimate, we use the estimate averaged from multiple frames to give a fine-grained estimate. We first theoretically propose the expression of the estimator variance in a single frame. Then, we investigate how many frames are necessary to reduce the estimator variance to a sufficiently small value such that the averaged estimate can satisfy the required $\langle \alpha, \beta \rangle$ accuracy.

The second challenge is to minimize the time cost when passive RFID tags are used, on the premise that the required accuracy is guaranteed. In reality, the frame size f is usually less than 512, for practical reasons [70]. To make REB scalable to the large-scale RFID systems, we use the persistence probability p [70]. The reader initializes a frame with the size of f/p, but sends commands to terminate the frame after the first $f \leq 512$ slots. The settings of f and p are important to the performance of REB. Hence, we propose sufficient theoretical analysis to optimize the parameters f and p to minimize the time cost of REB.

The third challenge is to minimize the energy cost when the battery powered active tags are used. We also investigate the optimization of frame size f and persistence probability p to minimize the energy cost of REB. However, we find that the time cost and energy cost of REB cannot be minimized at the same time. When the energy cost is minimized, the execution time of REB can be quite long. Hence, we should jointly consider the time- and energy efficiency of REB instead of separately considering them. Finally, we reveal a trade-off between the time cost and energy cost, which can be flexibly adjusted by the protocol parameters.

4.1.6 *Novelty and advantage over prior art*

The key novelty of this chapter is in formulating the practically important problem of RFID estimation with the presence of a blocker tag and taking the first step towards an efficient solution. The key technical depth of this chapter is in proposing the unbiased estimator of genuine tags and addressing the three aforementioned technical challenges. The key advantage of REB over prior art is threefold: (1) REB is compliant to the EPC C1G2 RFID standard and does not require any modifications to off-the-shelf tags, it only needs to be implemented on readers as a software module; (2) Compared with the prior estimation protocols, REB jointly uses the number of persistent empty slots and the number of persistent singleton slots to eliminate the interference from the blocker tag, and thus, can correctly estimate the cardinality of genuine tags; (3) REB significantly outperforms the state-of-the-art identification protocols in terms of both time- and energy efficiency. For example, when $|U| = 50,000$ and the tag ratio $|B - G| : |B \cap G| : |G - B| = 1 : 1 : 1$, REB runs 178× faster than EDFSA [77] and 2,785× faster than TH [78], meanwhile revealing 281× and 333× improvement over EDFSA and TH in terms of energy efficiency, respectively.

The rest of this chapter is organized as follows. In Section 4.2, we describe the detailed design of REB, and give the functional estimator as well as the minimum frame number that can guarantee the required estimation accuracy. In Section 4.3, we propose rigorous analysis to optimize the involved parameters to minimize the time cost and energy cost of REB, respectively. In Section 4.4, we conduct extensive simulations to evaluate the performance of REB.

4.2   REB protocol

In this section, we first describe the system model used in this chapter. Then, an efficient *RFID estimation scheme with Blocker tags (REB)* is proposed to estimate the number of genuine tags by jointly using N_{00} and N_{11} observed in a time frame. We explicitly give the functional estimator, and point out that the estimation using a single time frame is hard to be accurate due to probabilistic variance. Hence, we propose to use multiple independent time frames to refine the estimation. This section finally presents rigorous theoretical analysis to investigate how many frames are needed to guarantee the desired estimation accuracy and how to avoid premature termination of REB.

4.2.1   System model

We consider the RFID system containing a single reader, a single blocker tag, and a population of genuine tags. The set of blocking IDs is represented by B, whose cardinality is b. The set of genuine tags is denoted as G, whose cardinality is g. We use U to denote the universal tag set, where $U = B \cup G$ and $|U| = u$. The IDs in $B - G$ do not correspond to any genuine tags, whose cardinality is denoted as b', i.e., $b' = |B - G|$.

The reader communicates with tags (including both genuine tags and virtual ones *simulated* by the blocker tag) under control of the backend server. The communication between the reader and tags are based on a time slotted way. Any two consecutive transmissions (from a tag to a reader or vice versa) are separated by a waiting time $\tau_w = 302\,\mu s$ [70]. According to the specification of the Philips I-Code system [79], the wireless transmission rate from a tag to a reader is 53 kb/s, that is, it takes a tag $\tau_t = 18.9\,\mu s$ to transmit 1 bit. The rate from a reader to a tag is 26.5 kb/s, that is, transmission of 1 bit to tags requires $\tau_r = 37.7\,\mu s$. Then, the time of a slot for transmitting m-bit information from a tag to the reader is $\tau_w + m \times \tau_t$; and the time of a slot for transmitting m-bit information from a reader to the tags is $\tau_w + m \times \tau_r$. The main notations used throughout the chapter are summarized in Table 4.1.

4.2.2   Protocol description

Our REB uses the standard framed-slotted Aloha protocol specified in EPC C1G2 [53] as the MAC layer communication mechanism. The reader initializes a slotted time frame by broadcasting a binary request $\langle R, f \rangle$, where R is a random number and f is the frame size (i.e., the number of slots in the forthcoming frame). Using the received parameters $\langle R, f \rangle$, each tag initializes its slot counter sc by calculating $sc = H(\text{ID}, R)$ mod f, and the hashing result follows a uniform distribution within $[0, f - 1]$. In many existing RFID literature [80–82], a widely accepted assumption is that a tag is capable of computing a seeded hash function. Moreover, Luo *et al.* have proposed a scheme to implement the seeded hash function in passive RFID tags with simple circuits [83]. When designing our REB protocol, we could obtain the specified hash function from the RFID manufacturer. The reader broadcasts a `QueryRep` command at the end of each slot. Upon receiving `QueryRep`, a tag decrements its slot counter

Table 4.1   Notations used in the chapter

Notations	Descriptions
$G / B / U$	Set of genuine tags; set of blocking IDs; union set
$g / b' / u$	$g = |G|; b' = |B - G|; u = |B \cup G|$
α / β	Required confidence interval; required reliability
$\hat{g}$	Estimate of g
f / p	Frame size; persistence probability
$E(\cdot) / \mathrm{Var}(\cdot)$	Expectation; variance
Z_β	The percentile of β. e.g., $Z_\beta = 1.96$ when $\beta = 95\%$
p_{00} / p_{11}	Probability that a slot pair is $\langle 0, 0 \rangle$; probability that a slot pair is $\langle 1, 1 \rangle$
N_{00} / N_{11}	# Of the persistent empty slots in a frame; # of the persistent singleton slots in a frame
$\mathscr{T} / \mathscr{E}$	Time cost of REB; energy cost of REB
ω	Energy cost on an active tag for transmitting RN16

sc by 1. In a slot, a tag will respond to the reader if its slot counter sc becomes 0. According to the occupation status, slots are classified into three types: *empty slot* in which no tag responds; *singleton slot* in which only one tag responds; *collision slot* in which two or more tags respond.

In the following, we present how our REB estimates the number of genuine tags by observing the slots in a frame. Since the backend server gets full knowledge of the simulated blocking IDs, it is able to predict which slots the blocking IDs are "mapped" to. Thus, it is able to construct a virtual ternary array $\mathbb{B}[0 \ldots f - 1]$. A bit in $\mathbb{B}[0 \ldots f - 1]$ is set to 0 when no blocking ID is mapped to this slot; 1 when only one blocking ID is mapped to this slot; c when two or more blocking IDs are mapped to this slot (a hashing collision). On the other hand, by observing the frame, the reader could get another array $\mathbb{BG}[0 \ldots f - 1]$, also consisting of f bits. A bit in $\mathbb{BG}[0 \ldots f - 1]$ is set to 0 when no tag responds in this slot; 1 when only one tag responds in this slot; c when two or more tags cause a collision in this slot. To distinguish a singleton slot from a collision one, each tag does not need to respond with the whole 96-bit ID. For efficiency, each tag responds with the RN16 (16-bit) [53] that is much shorter than the 96-bit tag ID. Two slots with the same index in $\mathbb{B}[0 \ldots f - 1]$ and $\mathbb{BG}[0 \ldots f - 1]$ are called a slot pair. In our scheme, the reader needs to record the numbers of the following two types of slot pairs.

- N_{00} is the number of *persistent empty* slot pairs $\langle 0, 0 \rangle$ (i.e., $\mathbb{B}[i] = 0\,AND\,\mathbb{BG}[i] = 0$, $i \in [0, f - 1)$).
- N_{11} is the number of *persistent singleton* slot pairs $\langle 1, 1 \rangle$ (i.e., $\mathbb{B}[i] = 1\,AND\,\mathbb{BG}[i] = 1$, $i \in [0, f - 1)$).

REB can estimate the cardinality of genuine tags by jointly using the number of persistent empty slots and that of persistent singleton slots. A persistent empty slot happens only when no ID in $U = B \cup G$ is mapped to this index. Thus, N_{00} reflects the cardinality of U (i.e., u). Later, we will show that a *monotonous* functional relationship can be established between u and N_{00}. REB uses this function to estimate u from N_{00}. Similarly, a persistent singleton slot happens when only one ID in $B - G$ is mapped to this index. Therefore, N_{11} reflects the cardinality $|B - G|$ (i.e., b'). Clearly, if we know u and b', we can get the cardinality g of genuine tags by calculating $g = u - b'$. It may not be sufficient to satisfy the required estimate accuracy by counting the numbers of N_{00} and N_{11} in a *single* frame. Hence, REB executes k independent frames with different random number R, and uses the averaged estimate as the fine-grained result.

Note that, the frame size should be set to no more than 512 in practice [70,78,84] (the detailed reasons can be found in [78]). If a large number of tags contend for such a short frame, most slots will become collision slots. To scale to a large tag population, the reader uses a persistence probability $p \in (0, 1]$ to *virtually* extends the frame size f to f/p, but *actually* terminates the frame after the first f slots [70]. Fundamentally, each tag participates in the actual frame of f slots with a probability p.

4.2.3 Functional estimator

In this section, we derive the functional estimator $\hat{g}$ from N_{00} and N_{11} for the REB protocol in one frame. For an arbitrary slot pair, the probability that it is $\langle 0, 0 \rangle$, denoted as p_{00}, is given as follows:

$$p_{00} = \left\{ 1 - \frac{p}{f} \right\}^u \approx e^{-\frac{up}{f}} \tag{4.1}$$

The approximation in (4.1) holds when f/p is relatively large [50,67,70]. The number of slot pairs $\langle 0, 0 \rangle$, i.e., N_{00}, follows *Bernoulli*(f, p_{00}). The expectation and variance of the variable N_{00} are presented as follows:

$$E(N_{00}) = f p_{00} = f e^{-\frac{up}{f}} \tag{4.2}$$

$$\mathrm{Var}(N_{00}) = f p_{00} \{1 - p_{00}\} = f e^{-\frac{up}{f}} \left\{ 1 - e^{-\frac{up}{f}} \right\} \tag{4.3}$$

Similarly, we use p_{11} to denote the probability that a slot pair is $\langle 1, 1 \rangle$, which is given as follows:

$$p_{11} = \binom{b'}{1} \left\{ \frac{p}{f} \right\} \left\{ 1 - \frac{p}{f} \right\}^{u-1} \approx \frac{b'p}{f} e^{-\frac{up}{f}} \tag{4.4}$$

The number of $\langle 1, 1 \rangle$ slot pairs, i.e., N_{11}, also follows *Bernoulli* (f, p_{11}). The expectation and variance of the variable N_{11} are presented as follows:

$$E(N_{11}) = f p_{11} = b'p e^{-\frac{up}{f}} \tag{4.5}$$

$$\mathrm{Var}(N_{11}) = f p_{11} \{1 - p_{11}\} = b'p e^{-\frac{up}{f}} \left\{ 1 - \frac{b'p}{f} e^{-\frac{up}{f}} \right\} \tag{4.6}$$

According to (4.2), u can be expressed as follows:

$$u = -\frac{f}{p} \ln \left\{ \frac{E(N_{00})}{f} \right\} \tag{4.7}$$

Dividing (4.5) by (4.2), we have:

$$\frac{E(N_{11})}{E(N_{00})} = \frac{b'p}{f} \Rightarrow b' = \frac{fE(N_{11})}{pE(N_{00})} \tag{4.8}$$

According to (4.7) and (4.8), g is expressed as follows:

$$g = u - b' = -\frac{f}{p} \ln\left\{\frac{E(N_{00})}{f}\right\} - \frac{fE(N_{11})}{pE(N_{00})} \tag{4.9}$$

By substituting N_{00} for $E(N_{00})$ and N_{11} for $E(N_{11})$ in (4.9), we get the estimator of g as follows:

$$\hat{g} = -\frac{f}{p} \ln\left\{\frac{N_{00}}{f}\right\} - \frac{fN_{11}}{pN_{00}} \tag{4.10}$$

The estimator in (4.10) specifies how to use the observed N_{00} and N_{11} to estimate the cardinality g of genuine tags.

4.2.4 Variance of estimator

The proposed estimator has an inherent variance due to the probabilistic nature of REB. The following lemma calculates the expression of the estimator variance.

Lemma 4.1. *Let f and p be the frame size and the persistence probability, respectively, b' be the size of $B - G$, and u be the size of $B \cup G$. The variance of the estimator is as follows:*

$$\mathrm{Var}(\hat{g}) = \frac{1}{fp^2} e^{\frac{up}{f}} \left(b'^2 p^2 + f^2 - b'fp\right) - \frac{f}{p^2} \tag{4.11}$$

Proof. According to (4.10), $\hat{g}$ is a function of N_{00} and N_{11}. Hence, we denote $\hat{g}$ as $\varphi(N_{00}, N_{11})$, that is, $\hat{g} = \varphi(N_{00}, N_{11})$. We present the Taylor's series expansion [85] of function $\varphi(N_{00}, N_{11})$ around (η_0, η_1), where $\eta_0 = E(N_{00})$ and $\eta_1 = E(N_{11})$.

$$\varphi(N_{00}, N_{11}) \approx \varphi(\eta_0, \eta_1) + (N_{00} - \eta_0)\frac{\partial \varphi}{\partial N_{00}} + (N_{11} - \eta_1)\frac{\partial \varphi}{\partial N_{11}} \tag{4.12}$$

We have the following equation by taking expectation of both sides of (4.12).

$$E\{\varphi(N_{00}, N_{11})\}$$
$$= \varphi(\eta_0, \eta_1) + \frac{\partial \varphi}{\partial N_{00}} E(N_{00} - \eta_0) + \frac{\partial \varphi}{\partial N_{11}} E(N_{11} - \eta_1) = g \tag{4.13}$$

Equation (4.13) infers that $\hat{g}$ is an unbiased estimator of g. In what follows, we calculate the variance of $\hat{g}$.

$$\mathrm{Var}(\hat{g}) = E\{\hat{g} - E(\hat{g})\}^2$$

$$= E\left\{(N_{00} - \eta_0)\frac{\partial\varphi}{\partial N_{00}} + (N_{11} - \eta_1)\frac{\partial\varphi}{\partial N_{11}}\right\}^2$$

$$= \mathrm{Var}(N_{00})\left\{\frac{\partial\varphi}{\partial N_{00}}\right\}^2 + \mathrm{Var}(N_{11})\left\{\frac{\partial\varphi}{\partial N_{11}}\right\}^2 + \tag{4.14}$$

$$2\mathrm{Cov}(N_{00}, N_{11})\left\{\frac{\partial\varphi}{\partial N_{00}}\right\}\left\{\frac{\partial\varphi}{\partial N_{11}}\right\}$$

As required by (4.14), we need to calculate the covariance $\mathrm{Cov}(N_{00}, N_{11}) = E(N_{00}N_{11}) - E(N_{00})E(N_{11})$, in which $E(N_{00}N_{11})$ is calculated as follows:

$$E(N_{00}N_{11}) = \sum_{x=0}^{f}\sum_{y=0}^{f-x} xy P\{N_{00} = x \wedge N_{11} = y\}$$

$$= \sum_{x=0}^{f}\sum_{y=0}^{f-x} xy \binom{f}{x}\{p_{00}\}^x \binom{f-x}{y}\{p_{11}\}^y \{1 - p_{00} - p_{11}\}^{f-x-y}$$

$$= p_{11}\sum_{x=1}^{f} f\{f - x\}\binom{f-1}{x-1}\{p_{00}\}^x \{1 - p_{00}\}^{f-x-1}$$

$$= \frac{p_{00}p_{11}f^2}{1 - p_{00}}\sum_{x=1}^{f}\binom{f-1}{x-1}\{p_{00}\}^{x-1}\{1 - p_{00}\}^{f-x} \tag{4.15}$$

$$- \frac{f\{f - 1\}\{p_{00}\}^2 p_{11}}{1 - p_{00}}\sum_{x=2}^{f}\binom{f-2}{x-2}\{p_{00}\}^{x-2}\{1 - p_{00}\}^{f-x}$$

$$- \frac{fp_{00}p_{11}}{1 - p_{00}}\sum_{x=1}^{f}\binom{f-1}{x-1}\{p_{00}\}^{x-1}\{1 - p_{00}\}^{f-x}$$

$$= \frac{p_{00}p_{11}f^2}{1 - p_{00}} - \frac{f\{f - 1\}\{p_{00}\}^2 p_{11}}{1 - p_{00}} - \frac{fp_{00}p_{11}}{1 - p_{00}}$$

$$= f\{f - 1\}p_{00}p_{11}$$

As also required by (4.14), we calculate the first-order partial derivatives of $\varphi(N_{00}, N_{11})$ as follows:

$$\frac{\partial\varphi}{\partial N_{00}}\Big|_{\substack{N_{00}=\eta_0 \\ N_{11}=\eta_1}} = e^{\frac{up}{f}}\left(\frac{b'}{f} - \frac{1}{p}\right)$$

$$\frac{\partial\varphi}{\partial N_{11}}\Big|_{\substack{N_{00}=\eta_0 \\ N_{11}=\eta_1}} = -\frac{1}{p}e^{\frac{up}{f}} \tag{4.16}$$

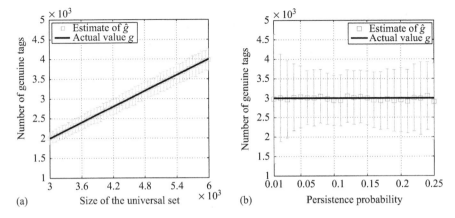

Figure 4.3   Comparing the estimates outputted by REB with the actual number of genuine tags. (a) Varying u = |U| from 3,000 to 6,000, where |B − G|:|B ∩ G|:|G − B| = 1:1:1. (b) Varying p from 0.01 to 0.25, where |B − G|=|B ∩ G|=|G − B|=2,000

We have obtained $E(N_{00}N_{11})$ in (4.15), $E(N_{00})$ in (4.2), and $E(N_{11})$ in (4.5). $Cov(N_{00}, N_{11})$ is calculated as follows:

$$Cov(N_{00}, N_{11}) = E(N_{00}N_{11}) - E(N_{00})E(N_{11}) = -b'pe^{-\frac{2up}{f}} \qquad (4.17)$$

By combining (4.3), (4.6), (4.16), (4.17) into (4.14), we then get the estimator variance shown in (4.11).   □

The simulation results in Figure 4.3 demonstrate that $\hat{g}$ in (4.10) is an unbiased estimator of the genuine tags. And the simulation results in Figure 4.4 reveal that the estimator variance observed from simulations match well with the theoretical value calculated by (4.11).

4.2.5   Refined estimation with k frames

Because of probabilistic variance, the estimate $\hat{g}$ got from a single frame is difficult to meet the predefined accuracy. By the law of large number [86], we issue k independent frames and use the average estimation result $\hat{g}_{\overline{k}} = \frac{1}{k}\sum_{j=1}^{k}\hat{g}_j$ to achieve a more accurate estimate in REB, where $\hat{g}_j$ is the estimate of g derived from the jth frame. We propose Theorem 4.1 to give the expression to determine if the frame number is adequate to ensure that REB achieves the required (α, β) accuracy.

Theorem 4.1. *Let α be the required confidence interval, β be the required reliability, f_j and p_j be the frame size and persistence probability used in the jth frame, respectively.*

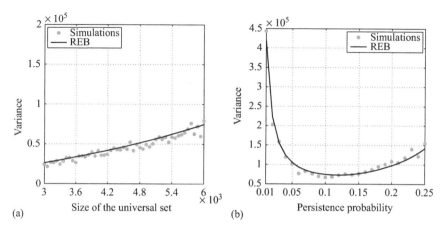

Figure 4.4   Comparing the estimator variance observed from simulations with the theoretical value calculated by (4.11). (a) Varying $u = |U|$ from 3,000 to 6,000, where $|B - G|:|B \cap G|:|G - B| = 1{:}1{:}1$. (b) Varying p from 0.01 to 0.25, where $|B - G|=|B \cap G|=|G - B|=2{,}000$

The average estimate $\hat{g}_{\overline{k}}$ obtained from k frames satisfies the accuracy requirement $P\left\{|\hat{g}_{\overline{k}} - g| \leq \alpha g\right\} \geq \beta$ when the frame number k satisfies the following inequality.

$$k \geq \frac{Z_\beta}{g\alpha}\sqrt{\sum_{j=1}^{k}\left\{\frac{1}{f_j p_j^2}e^{\frac{up_j}{f_j}}\left(b'^2 p_j^2 + f_j^2 - b'f_j p_j\right) - \frac{f_j}{p_j^2}\right\}} \tag{4.18}$$

Proof. We define $\hat{g}_{\overline{k}} = \frac{1}{k}\sum_{j=1}^{k}\hat{g}_j$ as the *average* estimate of k successive frames, where $\hat{g}_j$ is the estimate of the jth frame, $j \in [1, k]$. The reader initializes each frame with a different random seed. Hence, the estimate $\hat{g}_j$ is independent of each other. Thus, we have $E(\hat{g}_{\overline{k}}) = \frac{1}{k}\sum_{j=1}^{k}E(\hat{g}_j) = g$; and $\text{Var}(\hat{g}_{\overline{k}}) = \frac{1}{k^2}\sum_{j=1}^{k}\text{Var}(\hat{g}_j)$. Clearly, the average estimate $\hat{g}_{\overline{k}}$ still converges to the actual cardinality g. Given a required reliability β, the actual confidence interval is within $[g - Z_\beta\sqrt{\text{Var}(\hat{g}_{\overline{k}})}, g + Z_\beta\sqrt{\text{Var}(\hat{g}_{\overline{k}})}]$, where Z_β is a percentile of β, e.g., if $\beta = 95\%$, Z_β will be 1.96. To guarantee the required confidence α, we should guarantee:

$$\begin{cases} g + Z_\beta\sqrt{\text{Var}(\hat{g}_{\overline{k}})} \leq g + g\alpha \\ g - Z_\beta\sqrt{\text{Var}(\hat{g}_{\overline{k}})} \geq g - g\alpha \end{cases} \tag{4.19}$$

Substituting $\frac{1}{k^2}\sum_{j=1}^{k}\text{Var}(\hat{g}_j)$ for $\text{Var}(\hat{g}_{\overline{k}})$ and solving the above inequalities, we have:

$$k \geq \frac{Z_\beta}{g\alpha}\sqrt{\sum_{j=1}^{k}\text{Var}(\hat{g}_j)} \tag{4.20}$$

According to (4.11), we have $\mathrm{Var}(\hat{g}_j) = \frac{1}{f_j p_j^2} e^{\frac{up_j}{f_j}} (b'^2 p_j^2 + f_j^2 - b' f_j p_j) - \frac{f_j}{p_j^2}$. Substituting it into (4.20), we get the inequality in (4.18). $\qquad\qquad\square$

4.3 Parameter optimization

This section proposes rigorous theoretical analysis to optimize the values of frame size f and persistence probability p, to minimize the time cost and energy cost of REB, respectively. Then, we reveal a fundamental trade-off between the time cost and energy cost of REB, which can be flexibly adjusted by the system parameters. Through our analytical framework, we are able to configure the protocol parameters to achieve the desirable performance.

4.3.1 Minimizing time cost

Let $\mathscr{T}$ represents the total time cost of REB, t_v represents the time that the reader takes to transmit the parameters for frame initialization, t_μ represents the duration of each slot in the frame, and k represents the frame number required to satisfy the $\langle \alpha, \beta \rangle$ accuracy. Then, we have $\mathscr{T} = k \times (t_v + f \times t_\mu)$. We assume the values of f and p are consistently the same across all the frames. Thus, the required frame number k is transformed into

$$k = \frac{Z_\beta^2}{g^2 \alpha^2} \left\{ \frac{1}{fp^2} e^{\frac{up}{f}} (b'^2 p^2 + f^2 - b' fp) - \frac{f}{p^2} \right\} \qquad (4.21)$$

Then, the expression of the time cost $\mathscr{T}$ is as follows:

$$\mathscr{T} = \frac{Z_\beta^2 (t_v + f t_\mu)}{g^2 \alpha^2} \left\{ \frac{1}{fp^2} e^{\frac{up}{f}} (b'^2 p^2 + f^2 - b' fp) - \frac{f}{p^2} \right\} \qquad (4.22)$$

4.3.1.1 Optimizing p

Since f and p are correlated to minimize the total execution time, we first fix the value of f to get an optimized p. The following theorem calculates the optimal persistence probability p_{op} to minimize the time cost $\mathscr{T}$.

Theorem 4.2. *Given the sizes of $B \cup G$ and $B - G$ (i.e., u and b'), and the frame size f, if $\frac{\partial \mathscr{T}}{\partial p}|_{(p=1)} \geq 0$, we can solve the equation $e^{\frac{up}{f}} \left\{ \frac{ub'^2}{f^2} + \frac{u+b'}{p^2} - \frac{ub'}{fp} - \frac{2f}{p^3} \right\} + \frac{2f}{p^3} = 0$ to obtain the optimal persistence probability p_{op} to minimize the time cost $\mathscr{T}$. On the contrary, if $\frac{\partial \mathscr{T}}{\partial p}|_{(p=1)} < 0$, p_{op} should be 1.*

Proof. We calculate the expression of $\frac{\partial \mathscr{T}}{\partial p}$ as follows:

$$\frac{\partial \mathscr{T}}{\partial p} = \frac{Z_\beta^2 (t_v + f t_\mu)}{g^2 \alpha^2} \left\{ e^{\frac{up}{f}} \mathscr{D} + \frac{2f}{p^3} \right\}, \qquad (4.23)$$

where $\mathscr{D} = \frac{ub'^2}{f^2} + \frac{u+b'}{p^2} - \frac{ub'}{fp} - \frac{2f}{p^3}$. First of all, we prove that $\lim\limits_{p\to 0} \frac{\partial\mathscr{T}}{\partial p} < 0$. Then, we prove that the second-order derivative $\frac{\partial^2\mathscr{T}}{\partial p^2} > 0$, i.e., $\frac{\partial\mathscr{T}}{\partial p}$ is a monotonously increasing function of $p \in (0, 1]$. The corresponding proof can be found in our supplementary file or [87]. When $\frac{\partial\mathscr{T}}{\partial p}|_{(p=1)} \geq 0$, obviously, there is a value of $p_{op} \in (0, 1]$ that makes $\frac{\partial\mathscr{T}}{\partial p} = 0$. Moreover, we have $\frac{\partial\mathscr{T}}{\partial p} < 0$ when $p < p_{op}$; and $\frac{\partial\mathscr{T}}{\partial p} > 0$ when $p > p_{op}$. Thus, the time cost $\mathscr{T}$ achieves the minimum value when p_{op} is set to the value that satisfies the equation of $\frac{\partial\mathscr{T}}{\partial p} = 0$, by transforming which, we get the equation in theorem statement.

On the other hand, if $\frac{\partial\mathscr{T}}{\partial p}|_{(p=1)} < 0$, we assert that the first-order derivative $\frac{\partial\mathscr{T}}{\partial p}$ is always less than 0 for any value $p \in (0, 1]$. In this case, $\mathscr{T}$ is a monotonously decreasing function of $p \in (0, 1]$. Therefore, the optimal p_{op} should be set to 1.   □

4.3.1.2   Optimizing f

This section investigates the optimization of f to minimize the time cost $\mathscr{T}$. The following theorem reveals that we should directly set the frame size f to 512 in the large-scale RFID systems where the sizes of $|B \cup G|$ and $|B - G|$ are larger than 512. And in small-scale RFID systems, we will invoke an exhaustive search-based method to find the optimal frame size f. Fortunately, the range of f is an integer and is less than 512. Given a fixed frame size f, we can obtain the corresponding optimal p_{op} according to Theorem 4.2. By comparing all pairs of $\langle f, p_{op}\rangle$, we can obtain the optimal parameter pair that minimizes the time cost $\mathscr{T}$. Formally, we need to solve the following optimization problem.

$$\text{Minimize } \mathscr{T} = k \times (t_v + f \times t_\mu)$$

$$\text{s.t.} f \leq 512.$$

p is calculated by Theorem 4.2. $\qquad(4.24)$

$$k = \frac{Z_\beta^2}{g^2\alpha^2}\left\{\frac{1}{fp^2}e^{\frac{up}{f}}\left(b'^2p^2 + f^2 - b'fp\right) - \frac{f}{p^2}\right\}.$$

Theorem 4.3. *Given a large-scale RFID systems where the sizes of $B \cup G$ and $B - G$ are larger than 512 (i.e., $u > b' > 512$), we should set the frame size f_{op} to 512 for achieving the best time efficiency.*

Proof. Setting $p = 1$ in (4.23), we have $\frac{\partial\mathscr{T}}{\partial p}|_{(p=1)} = \frac{Z_\beta^2(t_v + ft_\mu)}{g^2\alpha^2}\left\{e^{\frac{u}{f}}\mathscr{M} + 2f\right\}$, where $\mathscr{M} = \frac{ub'^2}{f^2} + u + b' - \frac{ub'}{f} - 2f$. We first prove that $\lim\limits_{f\to 0}\{\frac{\partial\mathscr{T}}{\partial p}|_{(p=1)}\} > 0$ and $\lim\limits_{f\to +\infty}\{\frac{\partial\mathscr{T}}{\partial p}|_{(p=1)}\} < 0$ (refer to our supplementary file or [87]). Then, we prove that $\frac{\partial^2\mathscr{T}}{\partial p \partial f}|_{(p=1)} < 0$, which infers that $\frac{\partial\mathscr{T}}{\partial p}|_{(p=1)}$ is a monotonously decreasing function with respect to f (refer to our supplementary file or [87]). Hence, there exists a value of $f_\Delta > 0$ that makes $\frac{\partial\mathscr{T}}{\partial p}|_{(p=1,f=f_\Delta)} = 0$. And $\forall f_* \in (0, f_\Delta]$, we have $\frac{\partial\mathscr{T}}{\partial p}|_{(p=1,f=f_*)} \geq 0$.

According to Theorem 4.2, we could find an optimal persistence probability p_* corresponding to f_*, which satisfies the following equation.

$$e^{\frac{up_*}{f}}\left\{\frac{ub'^2}{f_*^2} + \frac{u+b'}{p_*^2} - \frac{ub'}{f_*p_*} - \frac{2f_*}{p_*^3}\right\} + \frac{2f}{p_*^3} = 0 \qquad (4.25)$$

According to (4.22), we get the time cost $\mathscr{T}_{(f_*,p_*)}$ when using frame size f_* and persistence probability p_*:

$$\mathscr{T}_{(f_*,p_*)} = t_v \times k_{(f_*,p_*)} + \frac{Z_\beta^2 t_\mu}{g^2\alpha^2} \times \psi, \qquad (4.26)$$

where $\psi = \frac{1}{p_*}e^{\frac{up_*}{f_*}}\left(b'^2p_*^2 + f_*^2 - b'fp_*\right) - \frac{f_*^2}{p_*^2}$. We calculate the first-order derivative $\frac{\partial\psi}{\partial f_*}$ as follows:

$$\frac{\partial\psi}{\partial f_*} = e^{\frac{up_*}{f_*}}\left\{-\frac{ub'^2p_*}{f_*^2} - \frac{u}{p_*} - \frac{b}{p_*} + \frac{ub'}{f_*} + \frac{2f}{p_*^2}\right\} - \frac{2f_*}{p_*^2} \qquad (4.27)$$

Consider the equation in (4.25), we find that $\frac{\partial\psi}{\partial f_*} = 0$. In Theorem 4.4, we have proved that $(gp\omega)k$ is a decreasing function of f. Hence, k is also a decreasing function of f, we then have $\frac{k_{(f_*,p_*)}}{\partial f_*} < 0$. Further, we have $\frac{\partial\mathscr{T}_{(f_*,p_*)}}{\partial f_*} = t_v \times \frac{k_{(f_*,p_*)}}{\partial f_*} + \frac{Z_\beta^2 t_\mu}{g^2\alpha^2} \times \frac{\partial\psi}{\partial f_*} < 0$. That is, the time cost $\mathscr{T}_{(f_*,p_*)}$ is a monotonously decreasing function of $f_* \in (0, f_\Delta]$.

Next, we use the method of reductio ad absurdum to prove that f_Δ is larger than 512 when $u > b' > 512$. If $f_\Delta < 512$ and $u > b' > 512$, we have $\mathscr{M} = \frac{ub'^2}{f^2} + u + b' - \frac{ub'}{f} - 2f > 0$. Then, $\frac{\partial\mathscr{T}}{\partial p}|_{(p=1)} = \frac{Z_\beta^2(t_v+ft_\mu)}{g^2\alpha^2}\left\{e^{\frac{u}{f}}\mathscr{M} + 2f\right\} = 0$ has no solution, which is an absurdum. Hence, when $u > b' > 512$, it is necessary that $f_\Delta > 512$. Recall that $\mathscr{T}_{(f_*,p_*)}$ is a monotonously decreasing function of $f_* \in (0, f_\Delta]$. Therefore, we should set f to 512 to minimize $\mathscr{T}$ in this case. $\qquad\square$

4.3.2 Minimizing energy cost

Energy cost is an important metric when designing an RFID application protocol for the systems where battery-powered active tags are used. This section investigates the impact of f and p on the energy efficiency of REB. Note that, we only take the energy consumption of genuine tags into consideration because recharging or replacing the batteries of a large number of active tags is seriously laborious. In contrary, the readers and blocker tag devices can be easily recharged because their number is normally limited, hence, we ignore the energy consumption of readers and the blocker tag devices. Following prior RFID literature [88] that also concerns the energy efficiency, we use the total number of tag transmissions to measure the total energy consumption, because transmitting data consumes much more power than receiving data. In an arbitrary frame, each of the g genuine tags has a probability p of responding RN16 to the reader. Therefore, the total power consumed by the genuine tags across k frames, denoted as $\mathscr{E}$, can be given by $\mathscr{E} = kgp\omega$, where ω represents the energy cost on an active tag for transmitting a response of RN16.

4.3.2.1   Optimizing f

The following theorem infers that we should set $f = 512$ to minimize the energy cost $\mathcal{E}$ of REB.

Theorem 4.4. *Given any value of the persistence probability $p \in (0, 1]$, we should set the frame size f to its maximum value of 512 to minimize the energy cost $\mathcal{E}$.*

Proof. We will prove that the required energy cost $\mathcal{E}$ is a monotonously decreasing function of the frame size f. The sufficient and necessary condition is that the first-order derivative of $\mathcal{E}$ with respect to f is always less than 0. Therefore, we calculate the first-order derivative of $\mathcal{E}$ with respect to f as $\frac{\partial \mathcal{E}}{\partial f} = \frac{z_\beta^2 \omega \mathcal{X}}{g \alpha^2 p}$, where $\mathcal{X}$ is given by

$$\mathcal{X} = e^{\frac{up}{f}} \left\{ 1 - \frac{up}{f} + \frac{(u - b')b'p^2}{f^2} - \frac{ub'^2 p^3}{f^3} \right\} - 1 \tag{4.28}$$

To prove $\frac{\partial \mathcal{E}}{\partial f} < 0$, we only need to prove $\mathcal{X} < 0$. Next, we will prove $\mathcal{X} < 0$ through two steps. First, we will prove $\mathcal{X}$ is a monotonically increasing function of the frame size f. Second, we will prove $\lim\limits_{f \to +\infty} \mathcal{X}$, i.e., the upper bound on $\mathcal{X}$, is always less than 0. Then, we can know that $\mathcal{X}$ is always less than 0 for any frame size $f \le 512$.

We calculate the expression of the first-order derivative of $\mathcal{X}$ with respect to f as follows. Additionally, we observe from (4.29) that $\frac{\partial \mathcal{X}}{\partial f}$ is always larger than 0. Hence, $\mathcal{X}$ achieves its largest value when $f \to +\infty$.

$$\frac{\partial \mathcal{X}}{\partial f} = e^{\frac{up}{f}} \left\{ \left(\frac{up}{\sqrt{2f^3}} - \frac{\sqrt{2}b'p}{\sqrt{f^3}} \right)^2 + \left(\frac{up}{\sqrt{2f^3}} - \frac{ub'p^2}{\sqrt{f^5}} \right)^2 \right.$$
$$\left. + \frac{4ub'^2 p^3}{f^4} + \frac{(\sqrt{2} - 1)u^2 b'p^3}{f^4} \right\} \tag{4.29}$$

In the following, we will prove that the *upper bound* on $\mathcal{X}$, i.e., the value of $\mathcal{X}$ when $f \to +\infty$, is always less than 0. Consider the expression of $\mathcal{X}$. For simplicity, we denote $1 - \frac{up}{f} + \frac{(u - b')b'p^2}{f^2} - \frac{ub'^2 p^3}{f^3}$ as $\mathcal{Q}$. Thus, $\mathcal{X} = e^{\frac{up}{f}} \mathcal{Q} - 1$. To prove $\mathcal{X} < 0$, we only need to prove $\mathcal{Q} < e^{-\frac{up}{f}}$. Using the Taylor series expansion, we have:

$$e^{-\frac{up}{f}} = 1 - \frac{up}{f} + \frac{(up)^2}{2!f^2} - \frac{(up)^3}{3!f^3} + \frac{(up)^4}{4!f^4} - \frac{(up)^5}{5!f^5}$$
$$+ \cdots + \left\{ \frac{(up)^\lambda}{\lambda!f^\lambda} - \frac{(up)^{\lambda+1}}{(\lambda+1)!f^{\lambda+1}} \right\} + \cdots, \tag{4.30}$$

where $\lambda = 6, 8, 10, \ldots$. Consider the tail item pair of (4.30). Since $\left| \frac{(up)^\lambda}{\lambda!f^\lambda} \right| \Big/ \left| -\frac{(up)^{\lambda+1}}{(\lambda+1)!f^{\lambda+1}} \right| = \frac{(\lambda+1)f}{up} > 1$ when $f \to +\infty$, we have $\left\{ \frac{(up)^\lambda}{\lambda!f^\lambda} - \frac{(up)^{\lambda+1}}{(\lambda+1)!f^{\lambda+1}} \right\} > 0$. Therefore, we have

$\mathscr{Y} = 1 - \frac{up}{f} + \frac{(up)^2}{2!f^2} - \frac{(up)^3}{3!f^3} + \frac{(up)^4}{4!f^4} - \frac{(up)^5}{5!f^5} < e^{-\frac{up}{f}}$. To prove $\mathscr{Q} < e^{-\frac{up}{f}}$, we only need to prove $\mathscr{Q} < \mathscr{Y}$. Hence, we calculate the expression of $\mathscr{Y} - \mathscr{Q}$ as follows:

$$
\begin{aligned}
\mathscr{Y} - \mathscr{Q} &= \frac{u^2 p^2}{2f^2} - \frac{ub'p^2}{f^2} + \frac{b'^2 p^2}{f^2} + \frac{ub'^2 p^3}{f^3} - \frac{u^3 p^3}{6f^3} \\
&\quad + \frac{u^4 p^4}{30f^4} + \frac{u^4 p^4}{120f^4} - \frac{u^5 p^5}{120f^5} \\
&= \left(\frac{up}{2f} - \frac{b'p}{f} \right)^2 + \frac{ub'^2 p^3}{f^3} + \left(\frac{up}{2f} - \frac{u^2 p^2}{\sqrt{30}f^2} \right)^2 \\
&\quad + \left(\frac{u^3 p^3}{\sqrt{30}f^3} - \frac{u^3 p^3}{6f^3} \right) + \frac{u^4 p^4}{120f^4} \left(1 - \frac{up}{f} \right)
\end{aligned}
\tag{4.31}
$$

Since $\frac{up}{f}$ is less than 1 when $f \to +\infty$, the expression of $\mathscr{Y} - \mathscr{Q}$ in (4.31) is always larger than 0. Then, we have $\mathscr{Q} < \mathscr{Y} < e^{-\frac{up}{f}}$. Obviously, we have $\mathscr{X} = e^{\frac{up}{f}}$ $\mathscr{Q} - 1 < 0$. Accordingly, the first-order derivative $\frac{\partial \mathscr{E}}{\partial f} = \frac{Z_\beta^2 \omega \mathscr{X}}{g\alpha^2 p} < 0$. Hence, for any fixed value of p, we should set f to its maximum value (i.e., 512 in practice), to minimize the energy cost. $\qquad\square$

4.3.2.2   Optimizing p

The following theorem infers that the energy cost $\mathscr{E} = kgp\omega$ is a monotonously increasing function of the persistence probability p. Hence, we should set the persistence probability $p \in (0, 1]$ as small as possible to decrease the energy cost $\mathscr{E}$.

Theorem 4.5. *Given any value of the frame size f, the energy cost $\mathscr{E} = kgp\omega$ is a monotonously increasing function with respect to the persistence probability p.*

Proof. We first prove that the second-order derivative $\frac{\partial^2 \mathscr{E}}{\partial p^2}$ is always larger than 0. Then, we can know the first-order derivative $\frac{\partial \mathscr{E}}{\partial p}$ is a monotonously increasing function of p. Thus, $\frac{\partial \mathscr{E}}{\partial p}$ gets close to its lower bound when $p \to 0$. Further, we prove that the lower bound $\lim\limits_{p \to 0} \frac{\partial \mathscr{E}}{\partial p}$ is always larger than 0. Then, we can assert that $\frac{\partial \mathscr{E}}{\partial p}$ is larger than 0 for any value of $p \in (0, 1]$. Specifically, we calculate the second-order derivative of $\mathscr{E}$ as follows:

$$
\frac{\partial^2 \mathscr{E}}{\partial p^2} = \frac{Z_\beta^2 \omega}{g\alpha^2} \left\{ e^{\frac{up}{f}} \times \mathscr{I} - \frac{2f}{p^3} \right\},
\tag{4.32}
$$

where $\mathscr{I} = \frac{u^2 b'^2 p}{f^3} + \frac{2ub'^2 - u^2 b'}{f^2} + \frac{u^2}{fp} - \frac{2u}{p^2} + \frac{2f}{p^3}$, which can be transformed as follows:

$$
\mathscr{I} = \left\{ \frac{ub'\sqrt{p}}{f\sqrt{f}} - \frac{u}{2\sqrt{fp}} \right\}^2 + \frac{2ub'^2}{f^2} + \left\{ \frac{\sqrt{3}u}{2\sqrt{fp}} - \frac{\sqrt{2f}}{p\sqrt{p}} \right\}^2 + \frac{(\sqrt{6}-2)u}{p^2}
\tag{4.33}
$$

Using Taylor series expansion, we have $e^{\frac{up}{f}} > 1 + \frac{up}{f} + \frac{u^2p^2}{2f^2} + \frac{u^3p^3}{6f^3} + \frac{u^4p^4}{24f^4}$. Since (4.33) reveals that $\mathscr{I} > 0$, we have:

$$
\begin{aligned}
\frac{\partial^2 \mathscr{E}}{\partial p^2} &= \frac{Z_\beta^2 \omega}{g\alpha^2} \left\{ e^{\frac{up}{f}} \times \mathscr{I} - \frac{2f}{p^3} \right\} \\
&> \frac{Z_\beta^2 \omega}{g\alpha^2} \left\{ \left(1 + \frac{up}{f} + \frac{u^2p^2}{2f^2} + \frac{u^3p^3}{6f^3} + \frac{u^4p^4}{24f^4} \right) \times \mathscr{I} - \frac{2f}{p^3} \right\} \\
&= \frac{Z_\beta^2 \omega}{g\alpha^2} \left\{ \left(\frac{u^3 b'p^2 \sqrt{p}}{2\sqrt{6f^7}} - \frac{u^3 p \sqrt{p}}{4\sqrt{6f^5}} \right)^2 + \frac{u^6 p^3}{32f^5} + \frac{u^3}{12f^2} \right. \\
&\quad + \left(\frac{u^2 b'p^2 \sqrt{u}}{2f^3} - \frac{u^2 p \sqrt{u}}{6f^2} \right)^2 + \left(\frac{ub'p\sqrt{2u}}{f^2} - \frac{u\sqrt{u}}{2\sqrt{2f}} \right)^2 \\
&\quad + \left(\frac{b'\sqrt{2u}}{f} - \frac{u\sqrt{u}}{2\sqrt{2f}} \right)^2 + \left(\frac{u^2 b'p\sqrt{p}}{2\sqrt{f^5}} - \frac{u^2 \sqrt{p}}{2f\sqrt{f}} \right)^2 \\
&\quad \left. + \frac{7u^4 b'^2 p^3}{12f^5} + \frac{3u^2 b'^2 p}{f^3} + \frac{u^5 p^2}{18f^4} \right\} > 0
\end{aligned}
$$

(4.34)

Equation (4.34) indicates that $\frac{\partial^2 \mathscr{E}}{\partial p^2}$ is always larger than 0, hence, $\frac{\partial \mathscr{E}}{\partial p}$ is a monotonously increasing function of $p \in (0, 1]$. Therefore, $\frac{\partial \mathscr{E}}{\partial p}$ gets close to its lower bound when $p \to 0$.

Next, we prove the lower bound, i.e., $\lim_{p\to 0} \frac{\partial \mathscr{E}}{\partial p}$, is always less than 0. First, we calculate its expression as follows:

$$
\frac{\partial \mathscr{E}}{\partial p} = \frac{Z_\beta^2 \omega}{g\alpha^2} \left\{ e^{\frac{up}{f}} \left(\frac{ub'^2 p}{f^2} + \frac{u}{p} + \frac{b'^2}{f} - \frac{ub'}{f} - \frac{f}{p^2} \right) + \frac{f}{p^2} \right\}
$$

(4.35)

For clarity, we denote $\frac{ub'^2 p}{f^2} + \frac{u}{p} + \frac{b'^2}{f} - \frac{ub'}{f} - \frac{f}{p^2}$ as $\mathscr{L}$. Then, $\frac{\partial \mathscr{E}}{\partial p} = \frac{Z_\beta^2 \omega}{g\alpha^2} \left\{ e^{\frac{up}{f}} \mathscr{L} + \frac{f}{p^2} \right\}$. To prove $\frac{\partial \mathscr{E}}{\partial p} > 0$ when $p \to 0$, we only need to prove $\mathscr{L} > -\frac{f}{p^2} e^{-\frac{up}{f}}$ when $p \to 0$. Consider the tail pair of items in (4.30). Since $\left| \frac{(up)^\lambda}{\lambda! f^\lambda} \right| / \left| -\frac{(up)^{\lambda+1}}{(\lambda+1)! f^{\lambda+1}} \right| = \frac{(\lambda+1)f}{up} > 1$ when $p \to 0$, we have $\left\{ \frac{(up)^\lambda}{\lambda! f^\lambda} - \frac{(up)^{\lambda+1}}{(\lambda+1)! f^{\lambda+1}} \right\} > 0$. Therefore, we have $\mathscr{Y} = 1 - \frac{up}{f} + \frac{(up)^2}{2! f^2} - \frac{(up)^3}{3! f^3} + \frac{(up)^4}{4! f^4} - \frac{(up)^5}{5! f^5} < e^{-\frac{up}{f}}$. Then, we have $-\frac{f}{p^2} \mathscr{Y} > -\frac{f}{p^2} e^{-\frac{up}{f}}$. To prove $\mathscr{L} > -\frac{f}{p^2} e^{-\frac{up}{f}}$, we only need to prove $\mathscr{L} > -\frac{f}{p^2} \mathscr{Y}$. To this end, we calculate the expression of $\mathscr{L} + \frac{f}{p^2} \mathscr{Y}$ as follows:

$$
\begin{aligned}
\mathscr{L} + \frac{f}{p^2} \mathscr{Y} &= \frac{ub'^2 p}{f^2} + \left(\frac{b'}{\sqrt{f}} - \frac{u}{2\sqrt{f}} \right)^2 + \left(\frac{u}{2\sqrt{f}} - \frac{u^2 p}{\sqrt{30f^3}} \right)^2 \\
&\quad + \left(\frac{u^3 p}{\sqrt{30f^2}} - \frac{u^3 p}{6f^2} \right) + \frac{u^4 p^2}{120f^3} \left(1 - \frac{up}{f} \right)
\end{aligned}
$$

(4.36)

In (4.36), since $\frac{up}{f} < 1$ when $p \to 0$, we have $\mathscr{L} + \frac{f}{p^2}\mathscr{Y} > 0$. Then, we get $\mathscr{L} > -\frac{f}{p^2}\mathscr{Y} > -\frac{f}{p^2}e^{-\frac{up}{f}}$. Further, we have $\frac{\partial \mathscr{E}}{\partial p} = \frac{z_\beta^2 \omega}{g\alpha^2}\left\{e^{\frac{up}{f}}\mathscr{L} + \frac{f}{p^2}\right\} > 0$ when $p \to 0$. Since the lower bound $\lim\limits_{p \to 0}\frac{\partial \mathscr{E}}{\partial p}$ is always larger than 0, we have $\frac{\partial \mathscr{E}}{\partial p}$ is always larger than 0 for any value of $p \in (0, 1]$. That is, the energy cost $\mathscr{E}$ is a monotonously increasing function of p. $\qquad\qquad\square$

4.3.3 Trade-off between time cost and energy cost

When the active tags are used, we should jointly take time- and energy efficiency into consideration, instead of separately considering these two metrics. Theorems 4.3 and 4.4 indicate that we should set the frame size f to 512 for achieving the best time- and energy efficiency in large-scale RFID systems. Hence, without otherwise specification, we set the frame size to 512. In terms of persistence probability p, no value of p simultaneously minimizes the time cost and energy cost. Theorem 4.5 infers that we should set the value of p as small as possible to minimize the energy cost. However, as illustrated in Figure 4.5, the execution time of REB will be significantly long when p is too small. Hence, we should make the trade-off between time cost and energy cost by controlling the setting of p, thereby satisfying the twofold requirements on both time- and energy efficiency. Specifically, we first leverage Theorem 4.2 to obtain the optimal persistence probability p_{op} to minimize the time cost. Clearly, both time cost and energy cost increase as p increases within the range of $[p_{op}, 1]$. Hence, we adjust the persistence probability p within the range of $(0, p_{op}]$, while ignoring the range of $p \in (p_{op}, 1]$.

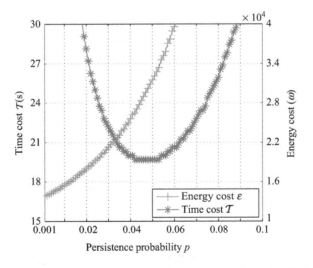

Figure 4.5   Trade-off between time cost and energy cost: $b = 10,000$, $b' = 5,000$, $g = 10,000$, $f = 512$, $\langle \alpha, \beta \rangle = \langle 0.02, 98\% \rangle$

4.3.4   Dynamic parameter optimization

Recall from the last two sections that we need to know the values of g, u and b' when calculating the optimal frame size f and persistence probability p. However, the actual values of g, u, and b' are not previously known, but are precisely what we want to estimate. Hence, we cannot calculate the optimal frame size f and persistence probability p at the start of REB.

For the first frame, we simply set the frame size f to 512 and set the persistence probability to $1/\hat{u}$, where the coarse value of $\hat{u}$ can be obtained by a fast method used in [70,75,78]. Specifically, the reader keeps issuing one-slot frames. The persistence probability follows a geometric distribution, $\frac{1}{2}$, $\frac{1}{4}$, $\frac{1}{8}$, ..., i.e., the persistence probability in the γth single-slot frame is $\frac{1}{2^\gamma}$. This process does not terminate until an empty slot appears. Assuming that the ℓth slot is the first empty slot, we have a coarse estimation of $\hat{u} = 1.2897 \times 2^\ell$ [72]. Note that, this process takes at most 32 slots in practice.

For the $(x+1)$th frame ($x \geq 1$), we could use the information observed from previous frames to estimate the values of g, u and b'. Equation (4.10) has given the estimator $\hat{g}$ of the genuine tags. According to Equations (4.7) and (4.8), we give the estimators for u and b' as: $\hat{u} = -\frac{f}{p}\ln\left\{\frac{N_{00}}{f}\right\}$ and $\hat{b'} = \frac{fN_{11}}{pN_{00}}$. Thus, we can use the temporary estimates averaged from the previous x frames, i.e., $\hat{g}_{\overline{x}} = \frac{1}{x}\sum_{j=1}^{x}\hat{g}_j$, $\hat{u}_{\overline{x}} = \frac{1}{x}\sum_{j=1}^{x}\hat{u}_j$, and $\hat{b'}_{\overline{x}} = \frac{1}{x}\sum_{j=1}^{x}\hat{b'}_j$, to calculate the optimal values of f and p to initialize the next frame. We have observed from the simulation results that REB can quickly converge to the near-optimal setting of f and p after a few frames.

4.3.5   Avoiding premature termination

After each frame, says the xth frame, REB gets $\hat{u}_{\overline{x}}$, $\hat{b'}_{\overline{x}}$, and $\hat{g}_{\overline{x}}$. However, their estimation is inaccurate due to the probabilistic variance. If we directly use them to calculate the R.H.S. of (4.18), the executed frame number x may have a chance to be larger than it, which is not true, and REB will have a premature termination (i.e., the currently achieved accuracy has not met the required one yet). In the following, we present how to avoid the premature termination.

Equation (4.11) has given the variance of estimator $\hat{g}$. Using the similar method in Section 4.2.4, we can calculate the variance of the estimators $\hat{u}$ and $\hat{b'}$ as follows:

$$\text{Var}(\hat{u}) = \frac{f}{p^2}\left(e^{\frac{up}{f}} - 1\right) \text{ and } \text{Var}(\hat{b'}) = e^{\frac{up}{f}}\left(\frac{b^2}{f} + \frac{b}{p}\right) \tag{4.37}$$

We use the average estimate $\hat{g}_{\overline{x}}$, $\hat{u}_{\overline{x}}$, and $\hat{b'}_{\overline{x}}$ got from previous x frames, and the frame size f_j as well as the persistence probability p_j used in the jth frame to calculate the estimation variance in the jth frame, i.e., $\text{Var}(\hat{g}_j)$, $\text{Var}(\hat{u}_j)$, and $\text{Var}(\hat{b'}_j)$, where $j \in [1, x]$. Then, we could get the variance of the average estimates, i.e., $\text{Var}(\hat{g}_{\overline{x}}) = \frac{1}{x^2}\sum_{j=1}^{x}\text{Var}(\hat{g}_j)$, $\text{Var}(\hat{u}_{\overline{x}}) = \frac{1}{x^2}\sum_{j=1}^{x}\text{Var}(\hat{u}_j)$, and $\text{Var}(\hat{b'}_{\overline{x}}) = \frac{1}{x^2}\sum_{j=1}^{x}\text{Var}(\hat{b'}_j)$.

To avoid the premature termination, when calculating the R.H.S. of (4.18), we use $\hat{u}_\uparrow = \hat{u}_{\overline{x}} + \delta\sqrt{\text{Var}(\hat{u}_{\overline{x}})}$ to substitute u, $\hat{b'}_\uparrow = \hat{b'}_{\overline{x}} + \delta\sqrt{\text{Var}(\hat{b'}_{\overline{x}})}$ to substitute the *first b'*, $\hat{b}'_\downarrow = \hat{b}'_{\bar{x}} - \delta\sqrt{\text{Var}(\hat{b}'_{\bar{x}})}$ to substitute the *second b'*, $\hat{g}_\uparrow = \hat{g}_{\bar{x}} + \delta\sqrt{\text{Var}(\hat{g}_{\bar{x}})}$ to substitute *g*. The *three-sigma rule* [89] indicates that $\delta = 3$ is large enough. In Section 4.4, simulation results demonstrate that this tactic can effectively avoid the premature termination.

4.4 Performance evaluation

In this section, we conduct extensive simulations to evaluate the performance of REB. Besides REB, we implemented two tag identification protocols including Tree Hopping (TH) protocol [78] and the classical enhanced dynamic framed-slotted Aloha (EDFSA) protocol [77]. Although no RFID estimation protocol can correctly estimate tag cardinality with the existence of blocker tags, we implemented several representative RFID estimation protocols and conduct comparison side-by-side to investigate the strength and weakness of our REB. The implemented RFID estimation protocols include PZE/PCE [50], FNEB [11], LoF [72], and ART [70]. Following many RFID literature [70,72,78], we assume that the communication channel is error free and a single reader covers all tags. We run each simulation 1,000 times and report the average results.

4.4.1 Verifying the convergence of f and p

The frame size f and persistence probability p that are actually picked by REB can approach to the near-optimal values after a few frames. The setting of parameters *f* and *p* is important to the performance of REB. To achieve the overall optimal *f* and *p*, it is necessary to know the values of *u*, *b'*, and *g* before the execution of REB, which, however, are what we want to estimate. As aforementioned in Section 4.3.4, we leverage the estimates of *u*, *b'*, and *g* obtained from previous frames to guide the parameter optimization of the next frame. Figure 4.6(a) plots the actual values of *f* and *p* used by each frame in a *small-scale* RFID system, where the optimal frame size *f* is 222 and the optimal persistence probability *p* is 1. On the contrary, Figure 4.6(b) plots the actual values of *f* and *p* used by each frame in a *large-scale* RFID system, where the optimal frame size *f* is 512 and the optimal persistence probability *p* is 0.1177. The simulation results reveal that REB can obtain the near-optimal settings of *f* and *p* after just a few frames.

4.4.2 Evaluating the actual reliability

The δ-sigma method proposed in Section 4.3.5 can effectively avoid the premature termination. REB (δ = 1) can always satisfy the required reliability. One of the most important performance metrics for estimation protocols is the actual reliability. In an arbitrary simulation, if the estimate $\hat{g}$ is within $[g(1 - \alpha), g(1 + \alpha)]$, we refer to it as a successful estimation. We record the *success times* among 1,000 independent simulations. The ratio, i.e., *success times*/1, 000, is treated as the *actual reliability*. Simulation results in Figure 4.7 reveal that REB ($\delta = 0$) does not always meet the

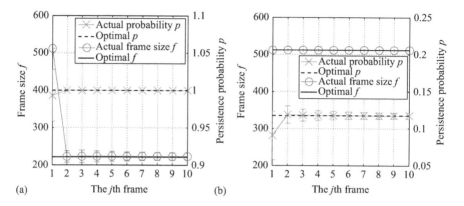

Figure 4.6   Verifying the convergence of f and p. α = 2%, β = 98%.
(a) Small-scale RFID system: |B − G| = 100, |B ∩ G| = 100,
|G − B| = 100; (b) large-scale RFID system: |B − G| = 2,000,
|B ∩ G| = 2,000, |G − B| = 2,000

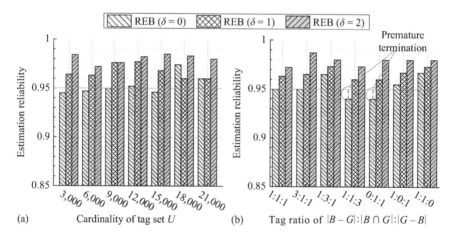

Figure 4.7   Evaluating the reliability of REB. α = 5%, β = 95%. (a) Tag ratio
|B − G|:|B ∩ G|:|G − B| is fixed to 1:1:1, and u varies from 3,000
to 21,000. (b) u is fixed to 9,000, and tag ratio varies

required reliability (i.e., $\beta = 95\%$). The reason lies in the variances if directly using $\hat{u}_{\overline{k}}$, $\hat{b}'_{\overline{k}}$, and $\hat{g}_{\overline{k}}$ to determine the termination condition. By taking their variances into consideration, the proposed δ-sigma-based termination tactic effectively avoids the premature termination. Simulation results in Figure 4.7 reveal that the actual reliability of REB ($\delta = 1$) and REB ($\delta = 2$) is always higher than the required one in various simulation environments.

4.4.3 Evaluating the time efficiency

Under the premise that the required estimation reliability is guaranteed, the most important metric is time efficiency. Recall that no existing estimation protocol can correctly approximate the cardinality of genuine tags in an RFID system with the presence of blocker tags. The only possible solution, to the best of our knowledge, is to perform the comprehensive identification protocols to identify the tags in the system. Hence, we compare REB with two representative identification protocols, i.e., the tree hopping (TH) protocol [78] and the enhanced dynamic framed-slotted Aloha (EDFSA) protocol [77]. TH protocol terminates after it traverses the whole tree. Normally, EDFSA repeats frames round by round until all the tags are identified and keep silent. However, with the presence of blocker tag, EDFSA cannot terminate by itself due to the continuous collisions caused by the tags in $B \cap G$. In frame of EDFSA, only the IDs in $(B - G) \cup (G - B)$ have the chance to be identified. We denote the set of identified IDs as S_{ident}. Since the reader does not know whether all IDs in $(B - G) \cup (G - B)$ are completely identified or even what percentage of them are identified. For the sake of EDFSA, we assume it can "intelligently" terminate once $\{|(B - G) \cup (G - B)| - |S_{\text{ident}}|\} < |G| \times \alpha$. In Section 4.4.2, the simulation results demonstrate that REB ($\delta = 1$) can always satisfy the required estimation accuracy. Hence, we set $\delta = 1$ in REB without otherwise specification, and use REB to denote REB ($\delta = 1$) for simplicity.

4.4.3.1 Impact of tag cardinality

REB significantly outperforms the state-of-the-art identification protocols, regardless of the tag population $|U|$. To investigate the impact of tag cardinality on the execution time of each protocol, we fix the tag ratio $|B - G|:|B \cap G|:|G - B|$ to 1:1:1, and vary u (indicating the system scale) from 20,000 to 50,000. The simulation results in Figure 4.8 demonstrate that our REB significantly outperforms HT and EDFSA. For example, when $u = 50,000$, REB runs about 178 times faster than EDFSA, and nearly 2,785 times faster than TH. Moreover, the execution time of TH and EDFSA grows linearly as u increases. In contrast, our REB has a stable execution time, which reveals its good scalability against tag cardinality u. The reason for the inefficiency of TH and EDFSA caused by the presence of blocker tag has been explained in Section 4.5.

4.4.3.2 Impact of tag ratio

REB significantly outperforms the state-of-the-art identification protocols in terms of time efficiency, regardless of the tag ratio $|B - G| : |B \cap G| : |G - B|$. The different tag ratio may have a significant impact on the execution time of protocols. Here, we fix $u = 30,000$ and evaluate the execution time of protocols with a varying tag ratio. The simulation results in Figure 4.9 demonstrate that the proposed REB protocol consistently runs hundreds of times faster than the state-of-the-art TH protocol and EDFSA protocol. Moreover, the results in Figure 4.9 clearly show the performance trend of the protocols with varying tag ratio.

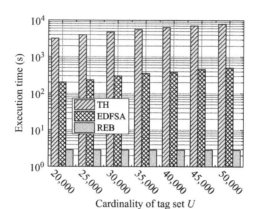

Figure 4.8   Evaluating the time efficiency of protocols with varying u. The tag ratio of |B − G|:|B ∩ G|:|G − B| is fixed to 1:1:1. α = 5%, β = 95%

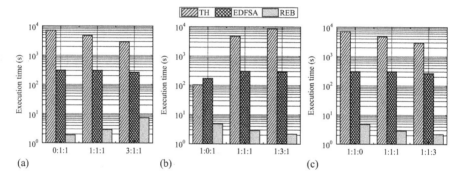

Figure 4.9   Evaluating the time efficiency of protocols with varying tag ratio. u is fixed to 30,000, and α = 5%, β = 95%. (a) Ratio of tags in B − G varies, (b) ratio of tags in B ∩ G varies, and (c) ratio of tags in G − B varies

Besides comparing with the tag identification protocols, we also conduct two set of simulations to compare our REB with the representative RFID estimation protocols including PZE/PCE [50], FNEB [11], LoF [72], and ART [70]. As shown in Figure 4.10, we compare REB with prior RFID estimation protocols without blocker tag. The simulation results reveal that our REB is indeed slower than several RFID estimation protocols. For comprehensive comparison, we also conduct simulations to evaluate the estimation accuracy of each protocol with varying number of blocking tag IDs. We observe from Figure 4.11 that all protocols can satisfy the required confidence interval $[g(1 − α), g(1 + α)]$ with a high reliability only without blocker tag (i.e., the number of blocking tag IDs is 0). However, as the number of blocking tag IDs increases, only our REB is able to correctly return the accuracy-guaranteed estimates. Experimental results show that other protocols significantly deviate from

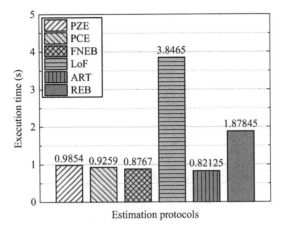

Figure 4.10   Evaluating the time efficiency of RFID estimation protocols without blocker tag. The number of tags is fixed to 10,000. $(\alpha, \beta) = (5\%, 95\%)$

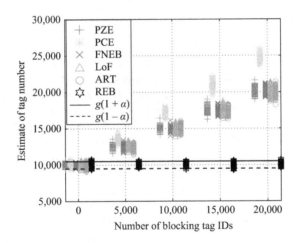

Figure 4.11   Evaluating the accuracy of RFID estimation protocols with varying number of blocking tag IDs. $|G| = 10,000$, $|B|$ varies from 0 to 20,000. $(\alpha, \beta) = (5\%, 95\%)$

the required confidence interval. All in all, our REB is the only protocol that can correctly perform RFID estimation with the presence of blocker tag.

4.4.4   Evaluating the energy efficiency

This section evaluates the energy efficiency of REB when the active RFID tags are used. Both TH and EDFSA only take time efficiency into consideration. For fair comparison, REB uses the values of f and p that minimizes the time cost.

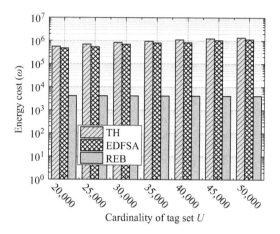

Figure 4.12 *Evaluating the energy efficiency of protocols with varying u. The tag ratio of $|B - G|$:$|B \cap G|$:$|G - B|$ is fixed to 1:1:1. $\alpha = 5\%$, $\beta = 95\%$*

4.4.4.1 Impact of tag cardinality

REB significantly outperforms the state-of-the-art identification protocols in terms of energy efficiency, regardless of the tag population $|U|$. In this set of simulations, we also fix the tag ratio of $|B - G|$: $|B \cap G|$: $|G - B|$ to 1:1:1 and vary the size of universal set U from 20,000 to 50,000. We make three main observations from the simulation results in Figure 4.12. First, the energy cost of TH and EDFSA increases as u (indicating the system scale) increases. The underlying reason is that TH and EDFSA aim at identifying all the tags; their energy cost will be proportional to the number of genuine tags. Since the tag ratio is kept to 1:1:1, the increase of u also indicates an increase of the number of genuine tags. Accordingly, the energy cost will increase. Second, the energy cost of REB is almost independent of the size of U. Third, our REB consistently outperforms TH and EDFSA in terms of energy efficiency. For example, when $u = 50,000$, the energy cost of TH and EDFSA is $1.38664 \times 10^6 \, \omega$ and $1.1686 \times 10^6 \, \omega$, respectively. And that of REB is just 4,158.47 ω, which represents 333 and 281 times improvement in terms of energy efficiency over TH and EDFSA, respectively.

4.4.4.2 Impact of tag ratio

REB significantly outperforms the state-of-the-art identification protocols in terms of energy efficiency, regardless of the tag ratio $|B - G|$: $|B \cap G|$: $|G - B|$. In this set of simulations, we fix the value of u and vary the tag ratio of $|B - G|$: $|B \cap G|$: $|G - B|$ to investigate the impact of tag ratio on the energy efficiency of each protocol. The simulation results shown in Figure 4.13 reveal that our REB consistently outperforms TH and EDFSA under different tag ratios by significantly reducing the energy cost.

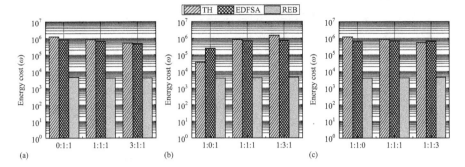

Figure 4.13 Evaluating the energy efficiency of protocols with varying tag ratio. u is fixed to 30,000, and $\alpha = 5\%$, $\beta = 95\%$. (a) Ratio of tags in $B - G$ varies, (b) ratio of tags in $B \cap G$ varies, and (c) ratio of tags in $G - B$ varies

4.4.5 Performance with constraints on time/energy cost

In practical applications, we may need to pose constraints on the time cost or energy cost of REB. For example, the trucks carrying tagged items may need to pass through a gate that deploys RFID readers within a short time window; to prolong the lifetime of active tags, we may need to pose a constraint on the energy cost of each execution of REB. Figure 4.14 plots the energy cost of REB with a given an upper bound on the time cost. With the simulation settings shown in Figure 4.14, the minimum time cost of REB is 15.7 s. When the required time threshold is within the range of [10, 14], we have to configure the parameters of REB to minimize its time cost. And thus, the energy cost of REB is stable at a certain level when the time threshold is within [10, 14]. On the contrary, when the time threshold is larger than 15.7 s, the energy cost of REB decreases as the time threshold increases. The underlying reason is that we can use a relatively small p when time threshold is large, and accordingly, the energy cost of REB decreases. Figure 4.15 plots the energy cost with a given constraint on the time cost. As the allowed energy cost increases, the execution time of REB decreases. When the constraint on energy exceeds $3 \times 10^4 \omega$, the constraint on energy will not be the bottleneck, then the time cost of REB remains at the same level.

4.5 Related work

In the infant stage of the study of RFID, a great deal of attention was paid to the problem of tag identification that aims to identify the exact tag IDs. Generally, there are two types of identification protocols: Aloha- [77] and Tree-based protocols [78]. Fundamentally, the Aloha-based protocol is a kind of Time Division Multiple Access (TDMA) mechanism. A tag ID can be successfully identified in a slot when only one tag responds in this slot. As for tree-based protocols, the reader broadcasts a 0/1 string to query the tags. A tag responds with its ID once it finds that the queried

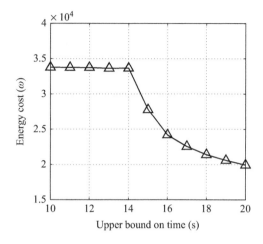

*Figure 4.14   Time cost of REB vs. given energy threshold. $u = 30,000$,
$|B - G|:|B \cap G|:|G - B| = 1:1:3$. $\alpha = 2\%$, $\beta = 98\%$*

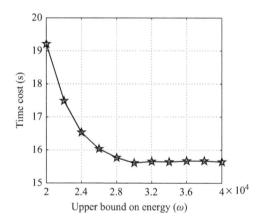

*Figure 4.15   Energy cost of REB vs. given time threshold. $u = 30,000$,
$|B - G|:|B \cap G|:|G - B| = 1:1:3$. $\alpha = 2\%$, $\beta = 98\%$*

string is the prefix of its ID. A reader identifies a tag ID when only one tag responds.
The execution time of identification protocols is proportional to the tag population
size. As explained in Section 4.1, the performance of tag identification protocols will
further deteriorate with the presence of blocker tags.

In recent years, a great effort has been made to study the problem of tag estimation
[50,54,67,70,72–75,82,84]. The RFID estimation protocols can be used for various
purposes. For the first example, when optimizing the performance of an Aloha-based
tag identification protocol, we should set the frame size to the number of tags, which,
however, is not known in prior. Here, an RFID estimation protocol that can quickly
and accurately tell the tag cardinality is desirable. For the second example, consider

the stock monitoring in a retailer, we only need to know the tag cardinality instead of the exact tag IDs. Hence, we prefer the fast RFID estimation protocols to the time-consuming tag identification protocols. To the best of our knowledge, the first piece of work concerning with RFID estimation was proposed by Kodialam *et al.* in 2006 [50]. They proposed PZE that uses the number of empty slots in a time frame, and PCE that uses the number of collision slots in a time frame to estimate the number of tags. The underlying reason is that they can establish a monotonous functional relationship between the number of tags and the number of empty slots or collision slots. Then, they can leverage the number of empty slots or collision slots observed from a time frame to perform the estimation of tag cardinality. Kodialam *et al.* have demonstrated that PZE and PCE are complementary to each other well, and combining them gives hybrid estimation protocol called UPE that performs well for a wide range of tag set cardinalities. Qian *et al.* [72] exploited the hashing with geometric distribution to estimate the cardinality of tags, and thus, proposed the Lottery Frame (LoF) scheme. In LoF, each tag has a probability of $\frac{1}{2^i}$ to choose the ith slot to respond to the reader. LoF requires the reader to distinguish empty slots from nonempty slots. An intuitive insight behind LoF's estimator is that the more tags are there in the system, the longer the length of continuous nonempty slots is expected to be. Hence, they can leverage the latter variable observed from a time frame to estimate the number of tags. Shahzad *et al.* proposed the Average Run based Tag estimation (ART) by observing the average length of sequences of consecutive nonempty slots [70]. ART is fast because its estimator has a smaller variance than previous RFID estimation protocols. Zheng *et al.* proposed zero-one estimator (ZOE) in which the responses from all the tags aggregate in the single time slot, and thus, allow ZOE to make extensive use of each time slot. Chen *et al.* proposed to put together various estimation schemes as building blocks in a proper manner, and thus achieve a more efficient protocol [73]. Li *et al.* argued that besides time efficiency, energy efficiency is also an important issue that must be carefully dealt with when battery-powered active tags are used. They proposed an estimation scheme called Maximum Likelihood Estimator (MLE) to take the energy efficiency into consideration [75]. For multicategory RFID systems, the literature [76] proposed to estimate tag cardinality in each category with a simultaneous manner; and [62,67] studied the top-k query problem (i.e., pinpointing the k largest categories). For dynamic RFID systems, literature [65,82] studied how to quickly estimate the number of absent tags and that of the remaining tags.

4.6 Conclusion

This chapter formally defines a new problem of RFID estimation with the presence of blocker tags and makes the first piece of effort that towards providing an efficient solution. The proposed *RFID Estimation scheme with Blocker tags (REB)* is compliant with the commodity EPC C1G2 standard, and does not require any modifications to off-the-shelf RFID tags. REB provides an unbiased functional estimator which can guarantee any degree of estimation accuracy specified by the users. Using REB, a retailer can monitor the product stock in a timely manner; meanwhile, the blocker tags are being used to protect the privacy of some important items. Rigorous analysis is given to optimize the parameters of REB to minimize its time cost and energy cost. A trade-off between the time cost and energy cost can be flexibly controlled to satisfy the practical requirements. Extensive simulation results has revealed the advantages of REB over prior schemes in terms of estimation accuracy, time efficiency, and energy efficiency.

Chapter 5

RFID detection—missing tags

5.1 Introduction

5.1.1 Background and motivation

Shoplifting, employee theft, and vendor fraud have become major causes of lost capital for retailers [93,96]. In 2011 alone, the retailers lost an estimated 34.5 billion dollars due to these causes [90]. With the benefits of not requiring a line-of-sight and low cost of tags (e.g., 5 cents per tag [24]), radio frequency identification (RFID) systems have been deployed for monitoring products by affixing them with cheap passive RFID tags and using RFID readers, which are given the IDs of the tags that are being monitored, to detect and identify any missing tags. A tag is a microchip with an antenna in a compact package that has limited computing power and communication range. There are two types of tags: (1) passive, which power up by harvesting the radio frequency energy from readers and have communication range often less than 20 ft; (2) active, which have their own power sources and have relatively longer communication range. A reader has a dedicated power source with significant computing power. It transmits queries to a set of tags and the tags respond over a shared wireless medium. In this chapter, we deal with both passive and active RFID tags.

5.1.2 Summary and limitations of prior art

There are two types of missing tag detection and identification protocols: *probabilistic* [83,97] and *deterministic* [94,95,98]. The probabilistic protocols are faster but only report the event that some tags are missing, without pinpointing exactly which ones. The deterministic protocols return IDs of all the missing tags but are comparatively slower. Both approaches have their merits. In fact, they are complementary to each other and should be used together. For example, a probabilistic protocol may be scheduled to execute frequently to detect if any tags are missing. Once a missing tag event is detected, a deterministic protocol may be invoked to identify the tags that are missing. Several probabilistic protocols such as TRP [97] and EMTD [83] and deterministic protocols such as IIP [94], MTI [98], and SFMTI [95] have been proposed.

There are two key limitations of existing protocols. The first limitation is that all existing protocols require a perfect environment that contains only the expected tags whose IDs are already known to the reader. However, in reality, tag populations often contain unexpected tags whose IDs are unknown. Here, we give three examples. For the first example, in airports where an airline company uses RFID readers to monitor baggage of its passengers, the tags of other airline's baggage, which are in the vicinity of this airline's readers, also respond to the queries of this airline's readers. For the second example, in a large warehouse rented to multiple tenants, one tenant's RFID readers receive responses from tags of other tenants. For the third example, in a retail store that uses RFID readers to monitor only expensive merchandize such as jewelry, the readers will receive responses from tags of inexpensive merchandize as well. Similar scenarios exist in other settings such as hospitals, prisons, and shopping malls. Existing protocols propose variations in standard frame-slotted Aloha protocol to speed up the detection and identification process. However, they cannot handle the presence of unexpected tags because unexpected tags fill up unexpected slots in Aloha frames resulting in unexpected false positives.

The second major limitation of existing protocols is that except TRP, none of these protocols is compliant with the EPCGlobal Class 1 Generation 2 (C1G2) RFID standard [53]. These protocols require the manufacturers to put random bit sequences in tags that the tags use to calculate specialized hash functions. They also require the tags to be able to receive and interpret "prevector" and/or "postvector" frames and use the information in those frames to select slots in Aloha frames. Such functionalities are not provisioned in the C1G2 standard because tags, especially the passive ones, do not have enough computational power to calculate complex hashes and process such pre- and postvector frames. It is important for an RFID protocol to be compliant with the C1G2 standard because the cheap commercially available off-the-shelf (COTS) RFID tags follow the C1G2 standard. A protocol that is not compliant with the C1G2 standard will require home brewed tags, which will not only cost more but will also work only in limited settings. For example, if an airline uses a protocol and tags that are noncompliant with the C1G2 standard, it will be able to track its luggage only at its home airport but not at the airports in rest of the world, which support only the C1G2 compliant tags.

5.1.3   *Problem statement and proposed approach*

Now, we formally define the missing tag detection and identification problem. Let $\mathbb{E}$ represent the set of IDs of the expected tags, i.e., the tags that are expected to be present in a population and need to be monitored. Let an unknown number of tags, m, out of these $|\mathbb{E}|$ tags be missing, where $0 \leq m \leq |\mathbb{E}|$. Let $\mathbb{E}_p$ be the set of IDs of the remaining $|\mathbb{E}| - m$ tags that are actually present in the population. Let $\mathbb{U}$ be the set of IDs of all the unexpected tags in the population that do not need to be monitored. We neither know exactly which IDs belong to sets $\mathbb{E}_p$ and $\mathbb{U}$ nor do we know their sizes, but we do know that $\mathbb{E}_p \subseteq \mathbb{E}$. Let T be a threshold on the number of missing tags. Our objective is to design a missing tag monitoring protocol using which *a set of readers should quickly detect a missing tag event with a probability $\geq \alpha$ whenever the number of missing tags m is greater than or equal to the threshold T, and if required, identify the IDs of the missing tags*. This probability α is called the required reliability and lies in the range $0 \leq \alpha < 1$. Additionally, a tag monitoring protocol should (1) be able to estimate the number of missing tags if the exact IDs of the missing tags are not required, (2) work in single as well as multiple readers environment, and (3) be compliant with the C1G2 standard.

For the problem of detecting and identifying missing tags in the presence of unexpected tags, there are three seemingly obvious solutions based on previous work. The first solution is to let RFID readers repeatedly execute a tag collection protocol to first collect IDs of all tags and then compare them with the IDs in set $\mathbb{E}$ to detect and identify missing tags. This solution works; however, it is too slow. For example, even using the fastest existing tag collection protocol TH [44], our performance evaluation results show that this solution is 14.3 times slower than our scheme. Delayed identification of missing RFID tags may have unbearable consequences such as giving thieves enough time to steal and run. The second solution is to first execute a tag collection protocol to get the IDs of unexpected tags and then repeatedly execute an existing RFID missing tag detection or identification protocol. This solution has two limitations. First, it is slow because the missing tag detection protocol will have to monitor the unexpected tags in addition to the expected tags. Second, the missing tag detection protocol will report a missing tag event even when some unexpected tags go missing, which is not the requirement. Furthermore, these solutions can't be used in settings where readers aren't allowed to read IDs in set $\mathbb{U}$ due to privacy reasons. An example of such a setting is the aforementioned multitenant warehouse, where one tenant may not permit readers of other tenants to read the IDs of its tags.

The third solution is to repeatedly execute a tag estimation protocol and look for a net change in the population size. The limitation of this solution is that if some expected tags go missing but an equal or greater number of unexpected tags joins the population, the estimation protocol cannot detect the missing tag event. Furthermore, missing tag detection protocols are much faster compared to estimation protocols due to the knowledge of set $\mathbb{E}$ [94,95,97].

In this chapter, we propose a new protocol called <u>R</u>FID monitoring protocol with <u>un</u>expected tags (RUN), the first protocol that can achieve required reliability in detecting and identifying missing tags when unexpected tags might be present in the population. RUN uses the frame-slotted Aloha protocol specified in the C1G2 standard as its MAC layer communication protocol. In Aloha protocol, the reader first tells the tags a frame size f and a random seed number R. Each tag within the transmission range of the reader then uses f, R, and its ID to select a slot in the frame by evaluating a hash function $h(f, R, ID)$ whose result is uniformly distributed in $[1, f]$. Each tag has a counter initialized with the slot number it chose to reply. After each slot, the reader first transmits an end of slot signal and then each tag decrements its counter by one. In any given slot, all the tags whose counters are equal to 1 respond with a pseudorandom sequence called RN16. If no tag replies in a slot, it is called an *empty slot*. If one or more tags reply in a slot, it is called a *nonempty slot*. As per the C1G2 standard, tags do not transmit their IDs unless the reader specifically asks them to do so. In RUN, the reader checks if a slot is empty or nonempty using the RN16 sequence and never asks the tags to transmit their IDs. This preserves the privacy in settings where a reader is not allowed to read IDs of tags in set $\mathbb{U}$. C1G2 standard provisions this functionality of not asking the tags for their IDs.

RUN can detect a missing tag event and if required, it can identify the tags that are missing. To distinguish between these two functionalities of RUN, we represent it by RUN_D when used for detection of missing tags and by RUN_I when used for identification of missing tags. In rest of this chapter, whenever we use the word RUN, it means that we are referring to both RUN_D and RUN_I. Next, we briefly describe how RUN_D detects a missing tag event and how RUN_I identifies the missing tags.

Missing tag detection: To detect if any tags are missing, RUN_D executes multiple Aloha frames with different seeds. In each frame, each tag uses the seed for that frame to select its slot. As RUN_D already knows the IDs of all tags in set $\mathbb{E}$, it precomputes which tags in $\mathbb{E}$ will select which slots in each frame. Thus, it knows which slots in the frames should be nonempty if all the tags in $\mathbb{E}$ are present in the population. When a reader executes a frame, RUN_D compares the response in each slot of that frame with the corresponding slot in the precomputed frame. If it finds that a particular precomputed slot was nonempty but the corresponding slot in the executed frame is empty, it stops and declares that some tags are missing. To minimize the effect of unexpected false positives and consequently the detection time, RUN_D estimates the size of $\mathbb{U}$ implicitly without running an extra estimation phase and uses this estimate to calculate optimal values of system parameters.

Missing tag identification: To identify exactly which tags are missing from the population of RFID tags, instead of stopping on encountering the first empty slot that was nonempty in the precomputed frames, RUN_I continues and executes v frames with different seeds. On each encounter with an empty slot in a frame that was nonempty in the corresponding precomputed frame, it marks all the tags in $\mathbb{E}$ that should have responded in that slot as absent. The value of v is such that in executing v frames, RUN_I is expected to identify all missing tags.

5.1.4   Technical challenges and solutions

There are four key technical challenges in detecting and identifying missing tags. The first technical challenge is to handle the presence of unexpected tags. Due to the presence of such tags, it is possible that a particular slot that RUN expected to be nonempty due to a specific tag in $\mathbb{E}$ actually turns out to be nonempty even though that specific tag in $\mathbb{E}$ was missing. To address this challenge, RUN executes multiple frames with different seeds, which reduces the effects of such unexpected *false positives*. We derive an expression to calculate the false-positive probability due to tags in $\mathbb{E}_p \cup \mathbb{U}$ and use that expression to calculate optimal values of frame sizes and the number of times the frames should be executed to mitigate the effects of false positives.

The second technical challenge is to estimate the number of unexpected tags $|\mathbb{U}|$ in the population of RFID tags, which is required to calculate the optimal values of system parameters such as frame size f and number of times n the frames should be executed. To address this challenge, RUN first precomputes which slots in each frame will the tags in $\mathbb{E}$ not select. Second, it executes the frames and sees how many of such slots turn out to be nonempty. The number of such slots that are nonempty in the executed frames is a function of $|\mathbb{U}|$ but is independent of $|\mathbb{E}|$ because we know from the precomputed frames that the tags in $\mathbb{E}$ never select these slots. Therefore, only the tags in $\mathbb{U}$ could have selected these slots. Thus, by observing the number of slots that are empty in the precomputed frames and nonempty in the executed frames, RUN estimates the number of unexpected tags $|\mathbb{U}|$ in the population of RFID tags. Note that RUN does not carry out a separate estimation phase to estimate the size of $\mathbb{U}$. It obtains the estimate while executing Aloha frames to detect and identify missing tags and thus, does not incur any extra time cost.

The third technical challenge is to achieve the required reliability in smallest possible time in case of detection of missing RFID tags. To address this challenge, we use the expression for false-positive probability to derive a "reliability condition," which, if satisfied by the values of frame size f and number of frames n, guarantees that RUN_D will achieve the required reliability. These values of frame size f and number of frames n ensure with probability α that there will be at least one slot in the n frames that is nonempty in the precomputed frames and empty in the executed frames when $m \geq T$. To minimize RUN_D's execution time, we express it in terms of f and n and minimize it under the constraint that f and n satisfy the reliability condition.

The fourth technical challenge is to identify all missing tags in smallest possible time in case of identification of missing RFID tags. To address this challenge, we derive an expression to calculate the number of missing tags that RUN_I identifies in each frame while incorporating the false positives. Using this expression, we derive a second expression to calculate the number of frames RUN_I needs to execute to identify all missing tags. We use this second expression to calculate total execution time of RUN_I and minimize it with respect to frame size f to get the optimal value of frame size. In addition to minimizing execution time, the optimal value of frame size f also ensures that each missing tag is expected to map to at least one such slot that is nonempty in a precomputed frame and empty in the corresponding executed frame.

5.1.5 Key novelty and advantages over prior art

The key novelty of this chapter is twofold. First, we identify the problem of detecting and identifying missing tags in the practical scenario where unexpected tags are present. Second, we propose RUN_D for detecting missing tags and RUN_I for identifying missing tags in the presence of unexpected tags. RUN has two key advantages over prior art. First, it achieves the required reliability in the presence of unexpected tags, whereas none of the existing protocols achieve the required reliability. We have extensively evaluated and compared RUN with four state-of-the-art missing tag detection protocols (TRP [97], IIP [94], MTI [98], and SFMTI [95]) in a variety of scenarios for a large range of tag population sizes. Among existing protocols, SFMTI achieves the highest reliability of 67%, whereas RUN_D achieves arbitrarily high reliability as per the requirement such as 99%. Similarly, SFMTI identifies up to 60% of missing tags in the presence of unexpected tags, whereas RUN_I identifies 100% of missing tags. Second, it is compliant with the C1G2 standard whereas existing protocols, except TRP, are not.

5.2 Related work

Several probabilistic [83,91,97] and deterministic [94,95,98,99] missing tag detection and identification protocols have been proposed. The common and major drawback of all of these protocols is that none of them handle unexpected tags and assume that the readers already know the IDs of all tags that can be present in the population.

5.2.1 Probabilistic protocols

Tan *et al.* proposed the first probabilistic protocol called TRP [97]. TRP pre-computes slots in a frame and compares them with the executed slots to detect missing tags. The difference with RUN, however, lies in that TRP does not consider false positives from unexpected tags because they assume that the reader already knows all the IDs in the population. In comparison, RUN does not know which slots will cause false positives due to the presence of unexpected tags, making the problem challenging. Furthermore, for large populations, TRP requires frame size that exceeds the C1G2 specified upper limit of 2^{15}, which is not possible in practical RFID systems. Among existing protocols, TRP is the only one that is compliant with the C1G2 standard as long as the frame size is below 2^{15}. Luo *et al.* proposed another probabilistic protocol called EMTD [83]. Unfortunately, this protocol is noncompliant with the C1G2 standard because it assumes the RFID tags to be intelligent with enough computing power to implement a hash ring and calculate hashes using that ring. Bu *et al.* proposed a suite of protocols, namely Cardiff, Cardiff+, and Divar, to anonymously detect missing tags [91]. Cardiff is essentially the same as the frame slotted Aloha protocol except that the tags never transmit their IDs, rather transmit only a 10-bit sequence. Cardiff counts the number of tags in the population and if the counted number of tags is less than the expected number of tags, it declares that some tags are missing. Cardiff+ slightly enhances the performance of Cardiff by increasing the frame size from equal to the tag population size to 1.2 to 1.4 times the tag population size. In Divar, the authors propose to use dynamic sequence spread spectrum (DSSS) to verify the intactness of an RFID population. Unfortunately, RFID tags in general and C1G2 compliant tags in particular, cannot implement a sophisticated scheme such as DSSS. Neither of the three protocols in [91] handle unexpected tags. None of the existing probabilistic protocols have been designed to work in multiple-reader environment.

5.2.2 Deterministic protocols

Li *et al.* proposed a suite of protocols in [94] out of which IIP performs the best. IIP is noncompliant with the C1G2 standard due to following three reasons. First, it requires tags to interpret prevector frames and reply to the reader queries as described in those frames. Second, it requires frame sizes greater than 2^{15} for large populations. Last, it requires manufacturers to insert a ring of random bits in tag memory at the time of manufacturing. IIP cannot handle multiple readers and requires frame sizes greater than 2^{15} for large populations. Zhang *et al.* proposed a deterministic protocol in [98], called MTI, that handles multiple readers but is also noncompliant with the C1G2 standard. Note that MTI is referred to as protocol 3 in [98]. The protocol is essentially a tag collection protocol that first collects the IDs of all the tags in the population and then checks which tags in the population are missing. The authors have not provided a theoretical frame work to calculate optimal system parameters, due to which MTI cannot be used to achieve an arbitrary desired accuracy. Liu *et al.* proposed a deterministic protocol in [95], called SFMTI. Although this protocol handles the use of multiple readers, it is noncompliant with the C1G2 tags because it requires tags to interpret nonstandardized vectors transmitted by the readers before and after selecting a slot in a frame. Zheng and Li proposed P-MTI, a physical layer missing tag identification protocol that leverages physical layer signals to identify missing tags [99]. Unfortunately, P-MTI works only when there are no unexpected tags because the reader in P-MTI needs the IDs of all tags in the population to precompute a sequence of bits for each tag. In the presence of unexpected tags with unknown IDs, the reader cannot compute such sequences, and thus not solve the system of linear equations, rendering P-MTI unable to identify missing tags.

5.3 System model

5.3.1 Architecture

For detecting and identifying missing tags, RUN uses a central controller connected with a set of readers that cover the area where the tags in set $\mathbb{E}$ are located. The use of a central controller ensures that all readers use consistent values of frame sizes and seeds when executing frames, which helps in efficiently aggregating and processing information returned by the readers. The readers use the standardized frame slotted Aloha protocol to communicate with tags and never ask the tags to transmit their IDs.

The use of multiple readers with overlapping coverage regions introduces following two problems: (1) scheduling the readers such that no two readers with overlapping regions transmit at the same time and (2) mitigating the effect of some tags responding to multiple readers due to overlap in the coverage region of those readers. For the first problem, the controller uses one of the several existing reader scheduling protocols [5,27,28,31] to avoid reader–reader collisions. For the second problem, RUN uses same values of frame size f and seed R across all readers with the help of the central controller. When a reader transmits seed R_i in its ith frame, it does not generate R_i on its own, rather it uses the ith seed R_i issued by the central controller. That is, each reader generates the same sequence of seeds in consecutive frames. As all readers use the same seed R_i in the ith frame, the slot number that a particular tag chooses in the ith frame of each reader covering this tag is the same i.e., $h(f, R_i, ID)$ evaluated by the tag results in same value for each reader. Once a reader completes its frame, it sends the responses to the central controller. The controller applies logical OR operator on all the ith frames from all readers and gets a single ith frame as if returned by one reader covering the entire tag population. It then compares it with the ith precomputed frame to detect and identify missing tags.

5.3.2 C1G2 compliance

RUN does not require any modifications to tags or readers. It only requires the readers to receive the frame size, persistence probability, and seed number from the controller and communicate the responses in the frames back to the controller. Persistence probability p is the probability with which a tag decides whether it will participate in a frame or not before selecting a slot in that frame. Later in the chapter, we will show how we use p to handle frame sizes that exceed the C1G2 specified upper limit of 2^{15}. As the C1G2 standard does not specify the use of p, COTS tags do not support it. To avoid making any modifications to tags, in RUN, the reader implements p by announcing a frame size of f/p but terminating the frame after the first f slots. To terminate a frame, the reader issues a SELECT command, specified in the C1G2 standard, with its *position*, *target*, and *action* parameters set to 0. This command "resets" all tags and they go into a state where they expect a new frame to start. For further details on frame termination, see Section 6 of [53].

5.3.3   Communication channel

We assume that the communication channel between readers and tags is reliable i.e., tags correctly receives queries from the readers and the readers correctly detect transmission of RN16 sequence in a slot if one or more tags in the population transmit in that slot. If the channel is unreliable, the solution proposed in [44] can be easily adapted for use with RUN.

5.3.4   Formal development assumption

To make the formal development tractable, we assume that instead of picking a single slot to transmit at the start of ith frame of size f, a tag independently decides to transmit in each slot of the frame with probability $1/f$ regardless of its decision about previous or forthcoming slots. Vogt first used this assumption for the analysis of Aloha protocol for RFID and justified its use by recognizing that this problem belongs to a class of problems called *occupancy problem*, which deals with the allocation of balls to urns [49]. Ever since, the use of this assumption has become a norm in the formal analysis of all Aloha-based RFID protocols [30,45,49].

The implication of this assumption is that a tag can end up choosing more than one slots in the same frame or even not choosing any at all, which is not in accordance with the C1G2 standard that requires a tag to pick exactly one slot in a frame. However, this assumption does not create any problems because the expected number of slots that a tag chooses in a frame is still one. The analysis with this assumption is, therefore, asymptotically the same as that without this assumption [51]. This independence assumption is made only to make the formal development tractable. In all our simulations, a tag chooses exactly one slot at the start of a frame. Table 5.1 lists the symbols used in this chapter.

5.4   Protocol for detection: RUN$_D$

To detect if any of the tags in set $\mathbb{E}$ is missing from the population, in RUN$_D$, the central controller executes up to n Aloha frames using the RFID readers. There are 6 steps involved in executing each frame. First, before executing any frame i, the controller

Table 5.1 Symbols used in the chapter

Symb.	Description
$\mathbb{E}$	Set of IDs of tags that need to be monitored
$\mathbb{E}_p$	Tags in $\mathbb{E}$ that are present in population
$\mathbb{U}$	Set of unexpected tags
$\mathbb{A}$	Set of IDs identified absent from population
$\mathbb{P}$	Set of IDs identified present in population
$\mathbb{I}_i$	Set of IDs of missing tags identified in ith frame
α	Required reliability
m	# Of tags in $\mathbb{E}$ missing from population
m_i	Expected # of missing tags not identified till start of ith frame
T	Threshold to detect missing tags
f	Frame size
$f_{\max}$	Max frame size allowed in the C1G2 $= 2^{15}$
n	# Of frames required to ensure detection
v	# Of frames required for identification
R_i	Seed used by the controller in ith frame
k_i	# Of slots that are 1 in the ith precomputed frame
$h(.)$	Hash function with output in $[1, f]$
p	Persistence probability
q	Prob. that at least one present tag selects a slot
g	Prob. that a missing tag event is detected
S_i	Total # of slots for RUN_I
S_d	Total # of slots for RUN_D
Z	Rand. var. for # of frames a tag participates in
D	Rand. var. for slot of first detection
X_j	Rand. var. for the event that jth slot is nonempty
$\mathcal{N}$	Rand var for # of nonempty slots in a frame
$\mathcal{N}_i^{01}$	Random var. for # of 0 slots in the ith precomputed frame that are 1 in the ith executed frame
X_{ij}	Random var. for the event that the jth 0 slot in the ith precomputed frame is 1 in the ith executed frame
Y_{ij}	Rand. var. for # of tags identified as missing in jth slot of ith frame
Y_i	Rand. var. for # of tags identified as missing in frame i

calculates the optimal values of frame size f_i, persistence probability p_i, and generates a random seed number R_i. Second, as the controller knows the IDs in set $\mathbb{E}$, it *precomputes* which tag in $\mathbb{E}$ will choose which slot in the ith frame. Thus, it knows which slots of the executed ith frame should be nonempty if all the tags in $\mathbb{E}$ were present and a single reader covered the entire population. It represents the nonempty slots in the precomputed frame with 1s and all other slots with 0s. Third, it provides each reader with the parameters f_i, p_i, and R_i and asks each of them to execute the ith frame using these parameters. The motivation behind using the same values of f_i, p_i, and R_i across all readers for the ith frame is to enable RUN_D to work with multiple readers with overlapping regions. As all readers use the same values of f_i,

p_i, and R_i in the ith frame, the slot number that a particular tag chooses in the ith frame of each reader covering this tag is the same i.e., $h(f_i/p_i, R_i, ID)$ evaluated by the tag results in same value for each reader. Fourth, each reader executes the frame on its turn as per the reader scheduling protocol and sends the responses in the frame back to the controller. Fifth, when the controller receives the ith frame of each reader, it applies logical OR operator on all the received ith frames and obtains a resultant ORed frame. This resultant ORed frame is same as if received by a single reader covering all the tags. Sixth, the controller compares all the slots in the precomputed ith frame with the corresponding slots in the resultant ORed ith frame. If there is any slot that is 1 in the precomputed frame but 0 in the resultant ORed frame, the controller detects this as a missing tag event because such a slot implies that all tags in $\mathbb{E}$ that are mapped to this slot in the precomputed frame are absent from the population. At this point, the controller stops the protocol and does not execute the remaining $n - i$ frames. If the controller does not detect a missing tag event even after each reader has executed n frames, it declares that the number of missing tags m is less than the threshold T.

5.5   Parameter optimization: RUN_D

Recall from the previous section that before executing any frame i, the controller calculates the optimal values of frame size f_i and persistence probability p_i. For this, the controller first estimates the value of number of unexpected tags $|\mathbb{U}|$ at the start of the ith frame, represented by $|\tilde{\mathbb{U}}_i|$, based on the responses from the tag population in the previous $i - 1$ frames. Then, using this estimate along with the values of number of expected tags $|\mathbb{E}|$, required reliability α, and threshold T, the controller calculates the optimal values of the frame size f_i and persistence probability p_i such that RUN_D achieves the required reliability in shortest time. Before asking the readers to execute the ith frame, the controller also recalculates the maximum number of frames that it should execute, represented by n_i. As the controller executes more and more frames, i.e., as i increases, the estimate $|\tilde{\mathbb{U}}_i|$ of the number of unexpected tags asymptotically becomes equal to the actual number of unexpected tags $|\mathbb{U}|$. Consequently, f_i, p_i, and n_i asymptotically become equal to constants f, p, and n, respectively. When the estimate of $|\mathbb{U}|$ does not change by more than 1% in 10 consecutive frames, the controller considers the estimate to be close enough to $|\mathbb{U}|$. At this point, the controller calculates the values of f_i, p_i, and n_i, puts $f = f_i$, $p = p_i$, and $n = n_i$, and uses these fixed values of f and p to execute subsequent frames until the total number of frames executed since the first frame become equal to n. Note that the controller executes n frames only if it does not detect any missing tag event in any frame. Otherwise, it terminates the protocol as soon as it detects a missing tag event. For the first frame, i.e., when $i = 1$, the controller uses $f_1 = 2 \times |\mathbb{E}|$, $p_1 = 1$, and $n_1 = \infty$. The choices of the values of f_1, p_1, and n_1 are arbitrary and do not really matter because as the controller executes more frames, the frame size, the persistence probability, and the number of frames converge to constants f, p, and n, respectively.

In rest of this section, we will derive equations that the controller uses at the start of each frame to calculate the optimal values of frame size f, number of times the frames should be repeated n, and persistence probability p to minimize the execution time of RUN_D while ensuring that its actual reliability is no less than the required reliability. We have dropped the subscript i from these parameters to make the presentation simple. To calculate these optimal values, the controller requires the estimate of $|\mathbb{U}|$. Next, we will first present a method to obtain this estimate at the start of any frame i based on the responses from the tag population in the previous $i - 1$ frames. Second, using the estimate of $|\mathbb{U}|$, we will derive an expression for the false-positive probability, i.e., the probability that a missing tag is detected as present. Third, we will use the expression for false-positive probability in conjunction with the required reliability α and threshold T to obtain an equation with three unknowns f, p, and n. To ensure that the actual reliability is greater than or equal to the required reliability, the controller must use the values of f, p, and n that satisfy this equation. We call this equation the *reliability condition*. Fourth, we will derive an expression for the total execution time of RUN_D and minimize it with respect to n to get an expression involving p and n. The controller simultaneously solves this expression with the reliability condition using $p = 1$ to obtain the optimal values of f and n. Last, we will show how to bring the value of f within limit when the optimal value of the frame size exceeds the C1G2 specified upper limit of 2^{15}, We will also calculate the expected number of slots RUN_D takes to detect the first missing tag event. Next, we describe these five steps in detail.

5.5.1 Estimating number of unexpected tags

In this section, we present a method to estimate the number of unexpected tags in the population at the start of any frame i. Although a lot of work has been done by the research community to estimate the number of tags present in an RFID tag population [45,50,54,92], there is no work on estimating the size of some subset of RFID tag population. In our case, that subset is the set of unexpected tags in the population.

Recall from Section 5.4 that in any frame i, the slots that are 0 in the ith precomputed frame are the slots that only the tags in set $\mathbb{U}$ can select when the reader executes the ith frame. This is because we have prior knowledge that the tags in set $\mathbb{E}$ will select only those slots in the ith executed frame that are 1 in the ith precomputed frame. The intuition behind our estimation method is that as the number of unexpected tags in a population increases, the number of slots that are 0 in a precomputed frame but are 1 in the corresponding executed resultant ORed frame also increase. The number of such slots in any given frame is a function of $|\mathbb{U}|$ and can, therefore, be used to estimate the value of $|\mathbb{U}|$.

Next, we derive an expression that relates the number of slots that are 0 in a precomputed frame but are 1 in the corresponding executed resultant frame with the value of $|\mathbb{U}|$. We will use this expression to obtain the estimate of $|\mathbb{U}|$. Let the size of the ith frame be f_i and let k_i of these f_i slots be 1s in the precomputed frames. Let j be the jth 0 slot in the precomputed frame. Thus, $1 \leq j \leq f_i - k_i$. Let X_{ij} be an indicator random variable for the event that the jth 0 slot in the ith precomputed frame turns out to be 1 in the ith executed resultant frame. The expected value of X_{ij} is given by

$$E[X_{ij}] = P\{X_{ij} = 1\} = 1 - \left(1 - \frac{p_i}{f_i}\right)^{|U|} \approx 1 - e^{-\frac{p_i}{f_i}|U|} \tag{5.1}$$

Let $\mathcal{N}_i^{01}$ be a random variable representing the number of slots that are 0 in the ith precomputed frame but 1 in the ith executed resultant frame. Thus, $\mathcal{N}_i^{01} = \sum_{j=1}^{f_i-k_i} X_{ij}$. As $\{X_{i1}, X_{i2}, \ldots, X_{i(f_i-k_i)}\}$ forms a set of identically distributed random variables, $E[\mathcal{N}_i^{01}]$ is given by

$$E[\mathcal{N}_i^{01}] = E\left[\sum_{j=1}^{f_i-k_i} X_{ij}\right] = (f_i - k_i) \times E[X_{ij}]$$
$$= (f_i - k_i) \times (1 - e^{-\frac{p_i}{f_i}|U|}) \tag{5.2}$$

Let $\tilde{\mathcal{N}}_i^{01}$ represent the observed value of the number of slots that were 0 in the ith precomputed frame but 1 in the corresponding executed resultant frame. Replacing $E[\mathcal{N}_i^{01}]$ in the equation above with $\tilde{\mathcal{N}}_i^{01}$ and solving for $|U|$ gives an estimate of $|U|$. This estimate is obtained by utilizing the information from the ith frame only. While this estimate may not be accurate, if we use the information from a large number of frames, the estimate will become more accurate. Specifically, we leverage the well-known statistical result that the variance in the observed value of a random variable reduces by x times if we take the average of x observations of that random variable. Therefore, to obtain the estimate $|\tilde{U}_i|$ of $|U|$ at the start of the ith frame, we obtain an estimate from each of the previous $i - 1$ frames and take their average. Solving (5.2) for $|U|$ and averaging over past $i - 1$ frames, the formal expression for $|\tilde{U}_i|$ becomes

$$|\tilde{U}_i| = -\frac{1}{i-1} \sum_{l=1}^{i-1} \frac{f_l}{p_l} \ln\left\{1 - \frac{\tilde{\mathcal{N}}_l^{01}}{f_l - k_l}\right\} \tag{5.3}$$

Finally, note that the controller obtains this estimate without executing any additional frames. It gets this estimate from the frames that it was already executing to detect missing tag events.

5.5.2 False-positive probability

A false positive occurs when all the slots that a particular missing tag maps to in the n precomputed frames turn out to be nonempty when the frames are executed because some other tags in the population also selected those slots. Lemma 5.1 gives the expression to calculate the false-positive probability.

Lemma 5.1. *Let m out of $|E|$ tags be missing, and let there be $|U|$ unexpected tags in the population. With persistence probability p, frame size f, and number of frames n, the false-positive probability, P_{fp}, is given by*

$$P_{fp} = \left\{1 - p\left(1 - \frac{p}{f}\right)^{|U|+|E|-m}\right\}^n \tag{5.4}$$

Proof. The total number of tags in the population are $|\mathbb{U}| + |\mathbb{E}| - m$. Consider an arbitrary tag in $\mathbb{E}$ that is missing from the population. As this tag participates in each precomputed frame with probability p, it is possible that it does not participate in one or more of the n precomputed frames. Let Z be the random variable for the number of precomputed frames in which this missing tag participates. Let q be the probability that a slot that this missing tag maps to in a precomputed frame is selected by one or more of the tags present in the population in the executed frame. Therefore,

$$P_{fp} = \sum_{z=0}^{n} P\{Z = z\} \times q^z \tag{5.5}$$

As a missing tag participates in each precomputed frame with probability p and there are n precomputed frames, the number of precomputed frames in which the missing tag participates follows a binomial distribution i.e., $Z \sim \text{Binom}(n, p)$. When a frame is executed, probability that a tag in the population chooses the same slot to which the missing tag maps in the precomputed frame is $\frac{p}{f}$. Probability that it does not choose that same slot is $1 - \frac{p}{f}$. Probability that none of the tags in the population choose that same slot is $(1 - \frac{p}{f})^{|\mathbb{U}|+|\mathbb{E}|-m}$. Probability that at least one tag in the population chooses that same slot is $1 - (1 - \frac{p}{f})^{|\mathbb{U}|+|\mathbb{E}|-m}$, which is the value of q. Therefore, (5.5) becomes

$$P_{fp} = \sum_{z=0}^{n} \binom{n}{z} p^z (1-p)^{n-z} \left\{ 1 - \left(1 - \frac{p}{f}\right)^{|\mathbb{U}|+|\mathbb{E}|-m} \right\}^z \tag{5.6}$$

To simplify the equation above, we use the binomial theorem, which states that

$$\sum_{z=0}^{n} \binom{n}{z} x^z y^{n-z} = (x+y)^n \tag{5.7}$$

Substituting $x = p \times \left\{ 1 - (1 - \frac{p}{f})^{|\mathbb{U}|+|\mathbb{E}|-m} \right\}$ and $y = 1 - p$ in the equation above, its left hand side (L.H.S.) becomes equal to the right hand side (R.H.S.) of (5.6) and its R.H.S. becomes equal to the R.H.S. of (5.4). Thus, the R.H.Ss of (5.4) and (5.6) are equal.   □

Figure 5.1 shows the theoretically calculated false-positive probability from (5.4) represented by the solid line and values of false-positive probability observed from simulations represented by the dots. To obtain this figure, we use $|\mathbb{E}| = 100$, $|\mathbb{U}| = 500$, $f = 300$, $p = 1$, and $n = 2$. Each dot represents the false-positive probability calculated from 100 runs of simulation. We observe that the theoretically calculated values match perfectly with the values observed from simulations, showing that our independence assumption that we stated in Section 5.3.4 does not cause the theoretical analysis to deviate from practically observed values. We also observe that as the number of missing tags increases, the false-positive probability decreases. This means that it is hardest for RUN_D to detect a missing tag event when $m = T$ and becomes easier as m increases beyond T. Thus, we will use $m = T$ in all further analytical development, because if RUN_D is able to detect a missing tag event with probability

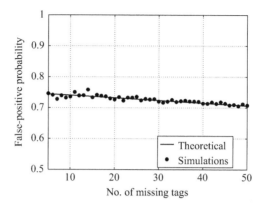

Figure 5.1   False-positive probability (P_{fp}) comparison

α when $m = T$, it will be able to detect a missing tag event with probability greater than α when $m > T$.

5.5.3 Achieving required reliability

Following theorem gives the *reliability condition* that the values of f, p, and n need to satisfy in order for $\mathrm{RUN_D}$ to be able to achieve the required reliability i.e., detect missing tag event with probability greater than or equal to α if the number of missing tags are greater than or equal to T.

Theorem 5.1. *Given a set $\mathbb{E}$ with expected IDs, set $\mathbb{U}$ with unexpected IDs, threshold T, and required reliability α, RUN_D will achieve the required reliability if the values of f, p, and n satisfy the reliability condition given below.*

$$f = \frac{p(T - |\mathbb{E}| - |\mathbb{U}|)}{\ln\left\{\frac{1-(1-\alpha)^{\frac{1}{nT}}}{p}\right\}} \tag{5.8}$$

Proof. The expression for P_{fp} in (5.4) gives the probability that $\mathrm{RUN_D}$ does not detect a given missing tag. Probability that it does not detect any of the T missing tags is P_{fp}^T. Probability that it detects at least one of the missing tags is $1 - P_{fp}^T$, which should be greater than or equal to α. In worst case, the probability that $\mathrm{RUN_D}$ detects at least one of the missing tags should at least be equal to α i.e., $1 - P_{fp}^T = \alpha$. Substituting P_{fp} by the R.H.S. of (5.4), we get

$$1 - \alpha = \left\{1 - p\left(1 - \frac{p}{f}\right)^{|\mathbb{U}|+|\mathbb{E}|-T}\right\}^{nT} \approx \left\{1 - pe^{-\frac{p}{f}(|\mathbb{U}|+|\mathbb{E}|-T)}\right\}^{nT} \tag{5.9}$$

Rearranging the equation above gives (5.8). □

5.5.4 Minimizing execution time

Following theorem gives the condition that p and n need to satisfy to make the execution time of RUN_D minimum under the constraint that it achieves the required reliability.

Theorem 5.2. *Given a threshold T and required reliability α, the execution time of RUN_D is minimum under the constraint that it achieves the required reliability if the values of p and n satisfy the following equation:*

$$p = \left\{ 1 - (1-\alpha)^{\frac{1}{nT}} \right\} \left\{ (1-\alpha)^{\frac{(1-\alpha)^{\frac{1}{nT}}}{nT(-1+(1-\alpha)^{\frac{1}{nT}})}} \right\} \tag{5.10}$$

Proof. Execution time is directly proportional to the total number of slots required to detect the missing tag event because the duration of each slot is the same and is equal to the time it takes the reader to distinguish between an empty and a nonempty slot (roughly around 300 μs for Philips I-Code RFID reader [56]). Let S_d represent the total number of slots, which equals the product of frame size f and total number of frames n, i.e., $S_d = f \times n$. To ensure that RUN_D achieves the required reliability, we use the value of f from (5.8). Thus,

$$S_d = \frac{pn(T - |\mathbb{E}| - |\mathbb{U}|)}{\ln\left\{ \frac{1-(1-\alpha)^{\frac{1}{nT}}}{p} \right\}} \tag{5.11}$$

Figure 5.2 plots S_d as a function of n. We observe that S_d is a convex function of n. Therefore, optimum value of n exists, represented by n_{op}, that minimizes the total number of slots S_d. To find optimal value of n, we differentiate (5.11) with respect to n and equate the resulting expression to 0, which gives (5.10). ☐

At the start of each frame, the controller replaces $|\mathbb{U}|$ with its estimate, puts $p = 1$ in (5.10) and solves it numerically using Brent's method to obtain the optimal value of number of frames n_{op}. Then, it puts $n = n_{op}$ and $p = 1$ in (5.8) to get the optimal value of frame size f_{op}. When the controller calculates f_{op} and n_{op} like this at the start of each frame, the execution time of RUN_D is minimized. At the same time, as the reliability condition is satisfied, the protocol achieves the required reliability.

5.5.5 Handling large frame sizes

For large populations, high required reliability, and/or small threshold, it is possible for the value of f_{op} to exceed the C1G2 specified upper limit of 2^{15}. Next, we describe how we use p to bring the frame size within limits. Bringing the frame size within limits comes at a cost of increased number of slots; greater than the minimum value of S_d that would have been achieved if the controller could use $f_{op} > 2^{15}$.

When we decrease the value of p, the number of tags that participate in a frame decrease. Therefore, intuitively, the required value of f should also decrease.

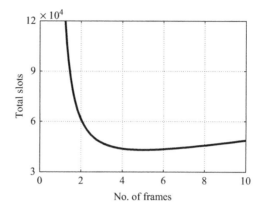

Figure 5.2   Number of slots (S_d) vs. number of frames (n)

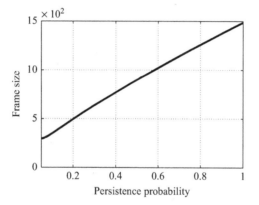

Figure 5.3   Frame size (f) vs. persistence probability (p)

Figure 5.3 confirms this intuition. This figure shows the plot of frame size vs. persistence probability, obtained using (5.8) and (5.10). We can see that when p decreases, f decreases. Participation by lesser tags means that participation by the tags belonging to both the sets $\mathbb{E}$ and $\mathbb{U}$ decreases. This increases the chances that a given missing tag will not map to any slot in a given precomputed frame, which means that chances of detecting its absence decrease. Therefore, the overall uncertainty in detection of missing tags increases. To reduce this uncertainty, intuitively, the value of n should increase when p decreases to achieve the required reliability. Figure 5.4 confirms this intuition. This figure shows the plot of number of frames vs. persistence probability, obtained using (5.8) and (5.10). We observe that when p decreases, n increases.

We use these two observations to reduce the value of f whenever $f_{op} > 2^{15}$. When $f_{op} > 2^{15}$, the controller uses $f = f_{max} = 2^{15}$ in (5.8), which leaves two unknowns, p and n, in the resulting equation. The controller solves the resulting equation simultaneously with (5.10) to get new values of p and n. The new value of p is less than 1

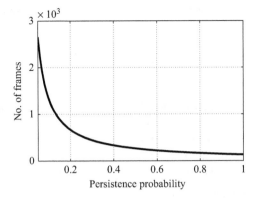

Figure 5.4 Number of frames (n) vs. persistence probability (p)

and the new value of n is greater than n_{op} because $f_{max} < f_{op}$. Putting $f = f_{max}$ in (5.8) and solving for n, we get

$$n = \frac{\ln\{1 - \alpha\}}{T \ln\left\{1 - pe^{\frac{p}{f_{max}}(T-|\mathbb{E}|-|\mathbb{U}|)}\right\}} \tag{5.12}$$

Replacing n in (5.10) with the R.H.S. of the equation above, and simplifying, we get

$$\frac{p^2(T - |\mathbb{E}| - |\mathbb{U}|)}{f(e^{\frac{p}{f_{max}}(|\mathbb{E}|+|\mathbb{U}|-T)} - p)} = \ln\left\{1 - pe^{\frac{p}{f_{max}}(T-|\mathbb{E}|-|\mathbb{U}|)}\right\} \tag{5.13}$$

The numerical solution of the equation above gives the new value of p, which the controller puts in (5.12) to get the new value of n. The controller uses these new values of n and p along with $f = f_{max}$ to precompute the ith frame. Although the total number of slots $S_d = f_{max} \times n > f_{op} \times n_{op}$, this is still the smallest under the constraints that the required reliability is achieved and the frame size does not exceed f_{max}.

5.5.6 Expected detection time

The values of f and n calculated above ensure that in executing n frames, RUN$_D$ will detect a missing tag event with probability $\geq \alpha$ if number of missing tags is $\geq T$. However, in many cases, the first missing tag event is detected before all n frames are executed. We calculate the expected value of the number of slots that RUN$_D$ takes to detect the first missing tag event. For this, we first calculate the probability that a missing tag event is detected in a given slot.

Lemma 5.2. *Given a set $\mathbb{E}$ with expected IDs, set $\mathbb{U}$ with unexpected IDs, and threshold T, when controller executes RUN$_D$ with persistence probability p and frame* size f, the probability g that a missing tag event is detected in any slot is given by the following equation.

$$g = \left\{1 - \left(1 - \frac{p}{f}\right)^T\right\} \times \left\{\left(1 - \frac{p}{f}\right)^{|U|+|E|-T}\right\} \qquad (5.14)$$

Proof. Probability that a missing tag event is detected in a given slot is the product of the probability that at least one missing tag maps to this slot in the precomputed frame and the probability that no tag in the population selects that slot in the executed frame. Considering the scenario where it is hardest for RUN_D to detect a missing tag event i.e., when $m = T$, probability that at least one of the missing tags maps to the given slot in the precomputed frame is $1 - \left(1 - \frac{p}{f}\right)^T$. The probability that none of the tags present in the population selects that slot is $\left(1 - \frac{p}{f}\right)^{|U|+|E|-T}$. The product of these two probabilities gives the expression for g in (5.14). □

Following theorem gives the expected value of the number of slots that RUN_D takes to detect the first missing tag.

Theorem 5.3. *Let D be the random variable for the slot number when the first missing tag event is detected. Given that the probability of detecting a missing tag event in a slot is g, as calculated in Lemma 5.2, frame size is f, and number of frames is n, we get*

$$E[D] = \frac{1 - (1 - g)^{fn} - fng(1 - g)^{fn}}{g} \qquad (5.15)$$

Proof. The random variable D follows geometric distribution with parameter g i.e., $P\{D = i\} = (1 - g)^{i-1}g$. The expected value, thus, becomes

$$E[D] = \sum_{i=1}^{S_d} iP\{D = i\} = \sum_{i=1}^{f \times n} ig(1 - g)^{i-1}$$

$$= \frac{1 - (1 - g)^{fn} - fng(1 - g)^{fn}}{g} \qquad (5.16)$$

□

Figure 5.5 shows the theoretically calculated expected slot number of first detection (calculated using (5.15)) represented by the solid line and values of slot number of first detection observed from simulations represented by the dots. This figure is obtained using $|\mathbb{E}| = 100$, $|\mathbb{U}| = 500$, and f, p, and n are calculated using the method explained in Section 5.5. Each dot represents average of results from 100 runs of simulation. We see that the values observed from simulations track the theoretically calculated values very well.

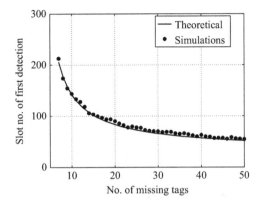

Figure 5.5   Comparison of analytical and empirical E[D]

5.5.7   Estimating number of missing tags

In some scenarios, exact IDs of missing tags may not be needed, rather only a count of the number of missing tags may be of interest. For example, consider a warehouse that stores a large number of boxes of shoes of the same type among other products. If some of those boxes of shoes are stolen (or may be shipped), the manager may only be interested in calculating *how many* boxes have left the warehouse. Furthermore, in such a scenario, if the number of boxes in the warehouse change frequently, estimation is a better choice compared to identification because estimation is usually much faster compared to identification for large populations [45,54].

RUN$_D$ can obtain a rough estimate of the number of tags missing from the population without incurring any extra time cost. We adapt the two most accurate and fast estimation schemes proposed until now i.e., enhanced zero-based estimator (EZB) [54] and average run-based tag estimator (ART) [45]. The idea behind EZB is that when the number of tags in the population increases, the expected number of empty slots in the frame decreases and the expected number of collision slots increases. A collision slot is a slot selected by two or more tags. Similarly, a singleton slot is a slot which is selected by only one tag. As we do not distinguish between singleton and collision slots, we adapt the idea behind EZB to our scenario, viz., as the tag population size increases, the number of empty slots decreases and the number of nonempty slots increases. Therefore, the number of empty slots and nonempty slots is a function of tag population size.

Next, we derive an expression to obtain a rough estimate of the tag population size. Let j be any arbitrary slot in the frame, where $1 \leq j \leq f$. Let X_j be an indicator random variable for the event that jth slot is nonempty, i.e., $X_j = 1$ if one or more tags in the population select the jth slot, otherwise $X_j = 0$. Let $\mathcal{N}$ be a random variable representing the number of nonempty slots in the frame. The value of $\mathcal{N}$ in terms of X_j is given by $\mathcal{N} = \sum_{j=1}^{f} X_j$. Therefore, when m tags are missing from the population

$$E[X_j] = P\{X_j = 1\} = 1 - \left(1 - \frac{p}{f}\right)^{|\mathbb{U}|+|\mathbb{E}|-m} \approx 1 - e^{-\frac{p}{f}(|\mathbb{U}|+|\mathbb{E}|-m)} \qquad (5.17)$$

As $\{X_1, X_2, \ldots, X_f\}$ forms a set of identically distributed random variables, $E[\mathcal{N}]$ is given by

$$E[\mathcal{N}] = E\left[\sum_{j=1}^{f} X_j\right] = f \times E[X_j] = f\left(1 - e^{-\frac{p}{f}(|\mathbb{U}|+|\mathbb{E}|-m)}\right) \qquad (5.18)$$

Solving the equation above for m, we get

$$m = \frac{p|\mathbb{U}| + p|\mathbb{E}| + f \ln\left\{1 - \frac{E[\mathcal{N}]}{f}\right\}}{p} \qquad (5.19)$$

To obtain an estimate $\tilde{m}$ of m, we execute n frames, count the number of 1s in each frame and take their average across the n frames. We replace $E[\mathcal{N}]$ in (5.19) with this average value and obtain the estimate $\tilde{m}$.

Similarly, we apply ART on the executed frames to obtain another estimate of m and take its average with $\tilde{m}$ obtained using (5.19) to get a final estimate of number of missing tags. The estimate of number of missing tags that we obtain this way is only a rough estimate. The accuracy of the estimate improves when more and more frames are executed. Note that this estimate is obtained without executing any additional frames.

5.6   Protocol for identification: RUN$_I$

In this section, we explain the RUN$_I$ protocol for *identifying* the tags in $\mathbb{E}$ that are missing from the population. RUN$_I$ protocol for identification builds on the RUN$_D$ protocol for detection. The controller maintains two sets $\mathbb{A}$ and $\mathbb{P}$ and initializes them as $\mathbb{A} = \phi$ and $\mathbb{P} = \mathbb{E}$. When the protocol completes, set $\mathbb{A}$ contains the IDs of all the tags in $\mathbb{E}$ that are missing from the population and set $\mathbb{P}$ contains the IDs of all the tags in $\mathbb{E}$ that are present. The controller calculates the values of v, p, and f, generates v random seed numbers R_i, where $1 \leq i \leq v$, and precomputes which tag will choose which slot in each frame. It executes the ith frame by providing each reader with f, p, and R_i. Each reader sends the responses in the ith frame back to the controller and the controller applies the logical OR operator on all received ith frames to obtain a resultant ORed frame. It then compares all the slots in the precomputed ith frame with the corresponding slots in the ORed ith frame. For each slot that is 1 in the precomputed frame but 0 in the ORed frame, the controller considers all the tags in $\mathbb{E}$ that mapped to this slot in the precomputed frame as absent. Let $\mathbb{I}_i$ represent the set of all the newly identified missing tags by the controller after comparing the ith ORed frame with the ith precomputed frame. After the ith frame, the controller adds the set $\mathbb{I}_i$ to set $\mathbb{A}$ and removes it from set $\mathbb{P}$ i.e., $\mathbb{A} = \mathbb{A} \cup \mathbb{I}_i$ and $\mathbb{P} = \mathbb{P} - \mathbb{I}_i$. The controller compares the frames and updates the sets $\mathbb{A}$ and $\mathbb{P}$ for all values of $i \in [1, v]$. RUN_I completes when the controller has compared all v frames. At this point, set $\mathbb{A}$ is expected to contain the IDs of all the missing tags.

5.7 Parameter optimization: RUN_I

In this section, we derive expressions to calculate the total number of frames v, persistence probability p, and frame size f that minimize the execution time of RUN_I while ensuring that after executing v frames, the controller is expected to have identified all tags in $\mathbb{E}$ that are missing from the population. For this, as a first step, we obtain an equation with three unknowns: v, p, and f. To ensure that RUN_I is expected to identify all missing tags, controller must use the values of v, p, and f that satisfy this equation. In the second step, we derive an expression for the execution time of RUN_I and minimize it with respect to p and f. This results in two more equations involving v, p, and f. Thus, there are three equations with three unknowns that the controller solves simultaneously to obtain the optimal values of v, p, and f. In the last step, the controller checks whether or not the frame size obtained from the simultaneous solution of these equations exceeds 2^{15}, and if it does, it brings its value in limits by adjusting the values of v and p in a similar way as described in Section 5.5.5.

5.7.1 Identifying all missing tags

Following theorem gives the equation that the values of v, p, and f must satisfy to ensure that the controller is expected to identify all missing tags in v frames.

Theorem 5.4. *Given a set $\mathbb{E}$ with expected IDs, set $\mathbb{U}$ with unexpected IDs, number of missing tags m, frame size f, and persistence probability p, the controller should execute v frames, calculated using the equation below, which will ensure that the expected number of missing tags it identifies in v frames equal the actual number of missing tags.*

$$v = 1 - \frac{\ln\{m\}}{\ln\left\{1 - p\left(1 - \frac{p}{f}\right)^{|\mathbb{U}|+|\mathbb{E}|-m}\right\}} \tag{5.20}$$

Proof. Let m_i be the expected number of missing tags that are not yet identified by the controller as missing at the start of ith frame. Let Y_{ij} be the random variable that represents the number of tags that the controller identifies as missing in jth slot of ith frame. Probability that the random variable Y_{ij} takes on a value l in jth slot of ith frame is the product of the probability that l out of the remaining m_i missing tags map to the jth slot of the precomputed ith frame and the probability that no tag in the population selects jth slot of the executed ith frame. Thus, the distribution of Y_{ij} is given by

$$P\{Y_{ij} = l\} = \binom{m_i}{l}\left(\frac{p}{f}\right)^l\left(1 - \frac{p}{f}\right)^{m_i-l} \times \left(1 - \frac{p}{f}\right)^{|\mathbb{U}|+|\mathbb{E}|-m} \tag{5.21}$$

Expected value of the number of tags that the controller identifies as missing in jth slot of the ith frame is, therefore, given by

$$E[Y_{ij}] = \sum_{l=1}^{m_i} lP\{Y_{ij} = l\}$$

$$= \sum_{l=1}^{m_i} l\binom{m_i}{l}\left(\frac{p}{f}\right)^l\left(1 - \frac{p}{f}\right)^{m_i-l} \times \left(1 - \frac{p}{f}\right)^{|\mathbb{U}|+|\mathbb{E}|-m}$$

$$= \frac{m_i p}{f} \times \left(1 - \frac{p}{f}\right)^{|\mathbb{U}|+|\mathbb{E}|-m} \tag{5.22}$$

The last step follows from the observation that the term on the left of "$\times$" is expectation of a binomial random variable $\sim \text{Binom}(m_i, \frac{p}{f})$.

Let Y_i be the random variable that represents the number of tags the controller identifies as missing in the ith frame. It is given by:

$$Y_i = \sum_{j=1}^{f} Y_{ij} \tag{5.23}$$

Expected value of Y_i, therefore, becomes:

$$E[Y_i] = \sum_{j=1}^{f} E[Y_{ij}] = \sum_{j=1}^{f} \frac{m_i p}{f}\left(1 - \frac{p}{f}\right)^{|\mathbb{U}|+|\mathbb{E}|-m}$$

$$= m_i p\left(1 - \frac{p}{f}\right)^{|\mathbb{U}|+|\mathbb{E}|-m} \tag{5.24}$$

After ith frame, the expected number of missing tags that the controller has not yet identified is m_{i+1}, and is given by:

$$m_{i+1} = m_i - E[Y_i] = m_i - m_i p\left(1 - \frac{p}{f}\right)^{|\mathbb{U}|+|\mathbb{E}|-m} \tag{5.25}$$

Equation (5.25) gives the recurrence relation to calculate expected number of missing tags the controller has not identified *after i* frames. Using this equation, we derive a closed form expression for m_i. As $m_1 = m$, it follows that:

$$m_i = m\left\{1 - p\left(1 - \frac{p}{f}\right)^{|\mathbb{U}|+|\mathbb{E}|-m}\right\}^{i-1} \tag{5.26}$$

Next, we calculate the value of v such that before the start of vth frame, the expected number of missing tags that the controller has not yet identified is only 1, which we expect to be identified in the vth frame. Thus, substituting 1 for m_i and v for i and solving for v, we get (5.20) in the theorem statement. □

5.7.2 Minimizing the execution time

Following theorem gives the two conditions that the values of p and f need to satisfy to make the execution time of RUN_I minimum under the constraint that the expected number of tags identified by the controller equal the actual number of missing tags.

Theorem 5.5. *The execution time of RUN_I is minimum under the constraint that the expected number of tags identified by the controller equal the actual number of missing tags if the values of p and f satisfy the following two equations:*

$$\frac{\partial}{\partial p}\left(f - \frac{f \times \ln\{m\}}{\ln\left\{1 - p(1 - \frac{p}{f})^{|\mathbb{U}|+|\mathbb{E}|-m}\right\}}\right) = 0 \tag{5.27}$$

$$\frac{\partial}{\partial f}\left(f - \frac{f \times \ln\{m\}}{\ln\left\{1 - p(1 - \frac{p}{f})^{|\mathbb{U}|+|\mathbb{E}|-m}\right\}}\right) = 0 \tag{5.28}$$

Proof. Let S_i represent the total number of slots that the controller executes for RUN_I. It is equal to the product of frame size f and total number of frames v, i.e., $S_i = f \times v$. To satisfy the constraint mentioned in the theorem statement, we use the value of v from (5.20) to calculate S_i. The total number of slots S_i is a concave function of p and f and therefore, optimum values of p and f exist, represented by p_{op} and f_{op}, that minimize the total number of slots S_i. Due to limitation of space, we have omitted the figures showing concavity of S_i with respect to p and f. To find optimal values of p and f, we take partial derivatives of S_i with respect to p and f and equate them to 0, which is what (5.27) and (5.28), respectively, show. □

The expanded expressions of (5.27) and (5.28) are long and complex and do not have a closed form solution. Therefore, we have omitted them to keep the presentation simple. The controller first numerically solves (5.27) and (5.28) using Levenberg–Marquardt algorithm to obtain the values of p_{op} and f_{op} and then substitutes these values in (5.20) to get the optimal value of v.

From (5.20), we see that to calculate the value of v, the controller needs the value of m, which is usually not known apriori. To address this problem, the controller starts with $m = |\mathbb{E}|$ because that results in the largest value of v and then, after each frame, estimates the value of m as explained in Section 5.5.7 and recalculates the value of v. As the controller executes more and more frames, the estimate of m becomes more and more accurate and eventually the value of v converges to its correct value. As the values of v are noisy in the initial few frames because the estimate of m is not accurate at the start, the controller uses some tolerance when deciding to stop i.e., it does not stop when it has obtained and compared v ORed frames, rather executes a few more frames to ensure that the value of v has stabilized. It stops only when the value of v has been relatively constant in last max $\{0.05v, 50\}$ frames, and it has already received at least v ORed frames. Figure 5.6 shows the evolution of the value of v in an example scenario as the controller executes frames. This figure is obtained

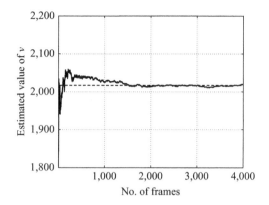

Figure 5.6   Convergence of value of v with frames

using $|\mathbb{E}| = 1,000$, $|\mathbb{U}| = 10,000$, and $m = 200$. We observe that the value of v is noisy at the start because the estimate of m is not accurate at the start, but as the controller executes more frames, the value converges to its correct value of 2017. In this example, it takes about 1,511 frames for v to converge to its correct value, which is much smaller than v itself.

5.8   Performance evaluation

We implemented RUN$_D$ and RUN$_I$ in MATLAB®. Although, none of the existing protocols handles the presence of unexpected tags and except for TRP, none of the existing protocols is C1G2 compliant, we still implemented four prior state of the art missing tag detection and identification protocols in MATLAB namely TRP [97], IIP [94], MTI [98], and SFMTI [95] and compared their performance with RUN$_D$ and RUN$_I$. We calculated parameter values for these protocols by following the instructions in their respective papers. We also implemented the fastest existing tag collection protocol TH [44]. We choose tag ID length of 64 bits as specified in the C1G2 standard. We do not make any assumptions about the distribution of IDs of expected, unexpected, and missing tags in the ID space because RUN$_D$ and RUN$_I$ are inherently independent of tag ID distributions. We calculate the optimal values of n, f, and p for RUN$_D$ as described in Sections 5.5.1–5.5.5, and of v, f, and p as described in Section 5.7.

In this section, we first evaluate the actual reliability of RUN$_D$ and the existing protocols for multiple values of required reliability, keeping the unexpected tag population size fixed and changing the number of missing tags. We also show the time taken by each protocol to detect first missing tag event. We further compare the number of missing tags identified by RUN$_I$ and existing missing tag identification protocols in $f \times v$ slots where f and v are calculated as described in Section 5.7. Second, we evaluate the actual reliability of RUN$_D$ and the existing protocols for multiple values of required reliability by keeping the number of missing tags fixed and changing the unexpected tag population size. We again show the time taken by each protocol to detect first missing tag event and the number of missing tags identified by RUN_I and existing protocols in $f \times v$ slots. Third, we study the actual reliability achieved by each protocol when the number of tags missing from the population is different from the value of threshold T. Fourth, we evaluate the estimation accuracy of RUN_D. Last, we compare the detection and identification times of our protocols with the fastest tag collection protocol TH. We also study whether there is a scenario in which the tag collection protocol is faster than our proposed protocols. For each scenario, we repeated simulations 500 times using different seeds and obtained the results from those 500 repetitions.

Recall that the controller in RUN_D uses (5.3) to estimate the number of unexpected tags $|\mathbb{U}|$ at the start of the ith frame. However, for the very first frame, the controller does not have any knowledge of $|\mathbb{U}|$ and thus does not know what frame size to start with. To address this problem, the controller executes a scheme based on Flajolet and Martin's probabilistic counting algorithm [9] that provides a quick but rough estimate of the number of tags in a population. This method for estimating the number of tags using Flajolet and Martin's algorithm was first proposed by Qian *et al.* in [23]. In this scheme, the reader executes multiple single-slot frames, where the persistence probability p follows a geometric distribution starting from $p = 1$ (i.e., $p = \frac{1}{2^{j-1}}$ in the jth single slot frame), until the reader gets an empty slot. As an example, if the number of tags in a population are 10,000, the expected value of the frame number in which reader will get an empty slot is $\lceil \log_2 10{,}000 \rceil = 14$. Suppose, the empty slot occurred in the jth single-slot frame, then $\tilde{t} = 1.2897 \times 2^{j-2}$ is a rough estimate of number of tags t in the population [9,23]. Elaborate details on how Flajolet and Martin's algorithm was adapted for use in RFID estimation can be found in [23]. The controller uses this value $\tilde{t}$ as the first very rough estimate of $|\mathbb{U}|$. This ensures that the size of the first frame is not grossly inaccurate, which in turns allows the controller to converge at the optimal frame size quickly. All results presented in this section incorporate this cost of obtaining this first rough estimate.

5.8.1 Impact of number of missing tags on RUN_D

RUN_D is the only protocol that achieves the required reliability in the presence of unexpected tags for any number of missing tags. Figure 5.7 shows the actual reliability achieved by RUN_D and all existing protocols for required reliability $\alpha = 0.9$ and 0.99, respectively. This figure is plotted using $|\mathbb{E}| = 1{,}000$, $|\mathbb{U}| = 10{,}000$ and m is varied from 50 to 900. The actual reliabilities are obtained using 100 runs of each protocol for each value of m. None of the existing protocols achieves the required reliability because none of them has been designed to handle unexpected tags. Among the existing protocols, SFMTI has the highest actual reliability of up to 0.67, which increases as the number of missing tags increase because increase in number of missing tags makes it easier for a protocol to detect a missing tag event.

RUN_D is the fastest protocol that achieves the required reliability compared to the existing protocols. Figure 5.8 shows the average times each protocol took to either

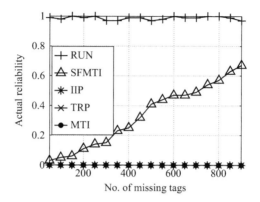

Figure 5.7   Actual reliability vs. # missing tags

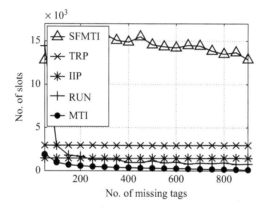

Figure 5.8   Detection time vs. # missing tags

detect the first missing tag event if it finds a missing tag or to complete execution if it does not find a missing tag. From this figure, MTI seems to have smaller detection time compared to RUN$_\mathrm{D}$, but when we analyze these figures in conjunction with Figure 5.7, we see that the actual reliability of MTI is close to 0, far lower than the required reliability. This shows that for majority of times, MTI completed execution without detecting any missing tags due to the presence of unexpected tags.

5.8.2   Impact of number of unexpected tags on RUN$_D$

RUN$_\mathrm{D}$ is the only protocol that achieves the required reliability in the presence of unexpected tags while existing protocols achieve the required reliability only when there are no unexpected tags in the population. Figure 5.9 shows the actual reliability obtained by RUN$_\mathrm{D}$ and the existing protocols for $\alpha = 0.9$ and 0.99, respectively. This figure is plotted using $|\mathbb{E}| = 1,000$, $m = 200$, and $|\mathbb{U}|$ is varied from 0 to 10,000. As before, the actual reliabilities are obtained using 100 runs of each protocol for

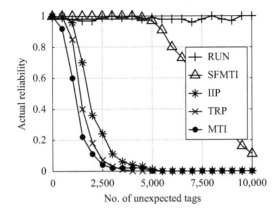

Figure 5.9   Actual reliability vs. # unexpected tags

each value of $|\mathbb{U}|$. RUN_D always achieves the required reliability whereas the existing protocols achieve the required reliability only when $|\mathbb{U}| = 0$, i.e., with no unexpected tags. As soon as the value of $|\mathbb{U}|$ increases, the actual reliability of existing protocols drops swiftly.

RUN_D *is the fastest protocol that achieves the required reliability compared to the existing protocols even when there are no unexpected tags in the population.* Figure 5.10 shows the average times each protocol took to either detect the first missing tag event or complete execution without detecting any missing tags. From this figure, MTI again seems to have smaller detection time compared to RUN_D when number of unexpected tags in the population is comparatively larger, but when we analyze these figures in conjunction with Figure 5.9, we see that actual reliability of MTI is close to 0 when number of unexpected tags in the population is comparatively larger.

5.8.3   Impact of number of missing tags on RUN_I

RUN_I *identifies up to 100% of all the missing tags in $f \times v$ slots, regardless of the number of missing tags, whereas the existing protocols do not identify more than 30% of all the missing tags in same number of slots.* This can be seen in Figure 5.11, which shows the bar plots of number of missing tags identified by each protocol for $m = 50$, 300, and 500. In Figure 5.8, which shows the average times each protocol took to detect the first missing tag event, we do not show a plot for RUN_I because RUN_I is same as RUN_D except that it does not stop after the first detection of missing tag event. Therefore, the average time RUN_I takes to detect the first missing tag event is same as that of RUN_D.

5.8.4   Impact of number of unexpected tags on RUN_I

RUN_I *identifies the largest number of missing tags in $f \times v$ slots among all the protocols even when $|\mathbb{U}| = 0$.* This can be seen in Figure 5.12, which shows the bar plots

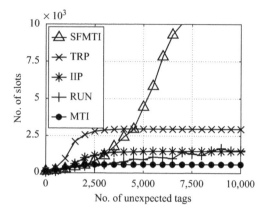

Figure 5.10   Detection time vs. # unexpected tags

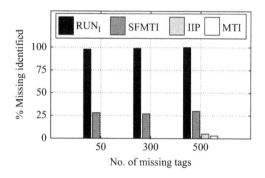

Figure 5.11   Percentage of identified missing tags vs. total number of missing tags

of number of missing tags identified by each protocol for $|\mathbb{U}| = 0, 3,000$, and $5,000$. RUN_I identifies up to 100% of all the missing tags, whereas the existing protocols do not identify more than 60% of all the missing tags. Note that Figures 5.11 and 5.12 do not compare RUN_I with TRP because TRP cannot *identify* the missing tags; it can only detect that some tags are missing.

5.8.5   Impact of deviation from threshold

The actual reliability of RUN_D exceeds the required reliability when the number of missing tags in the population exceed the threshold T, used to obtain the values of f, p, and n. This is seen in Figure 5.13, which plots the actual reliabilities of all protocols when number of missing tags are larger or smaller compared to T. This figure is made using $|\mathbb{E}| = 1,000$, $|\mathbb{U}| = 10,000$, $T = 200$, $\alpha = 0.99$, and m is varied from 50 to 900. Note that the actual reliability of RUN_D is less than the required reliability when the number of missing tags are less than T, but this is insignificant because as per the requirement, we are interested in detecting the missing tags only if the number of missing tags in a population exceed the threshold T.

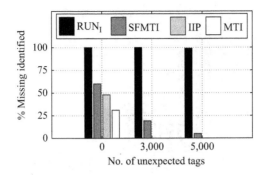

*Figure 5.12   Percentage of identified missing tags vs. total number of
unexpected tags*

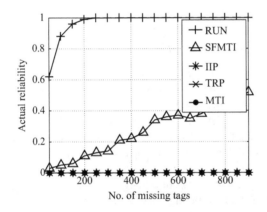

Figure 5.13   Effect of difference between m and T

5.8.6   Estimation accuracy

The estimates of number of missing tags obtained using the method described in Section 5.5.7 lie within ±5% of the actual number of missing tags when the required reliability $\alpha \geq 0.95$. The estimates are dependent on required reliability because required reliability is a factor that determines the number of frames n that the controller executes to ensure that RUN_D achieves the required reliability. The larger the value of the required reliability α, greater is the value of n and thus, more accurate are the estimates. Figure 5.14 plots the coefficient of variation of the estimates of number of missing tags obtained using the method described in Section 5.5.7. Coefficient of variation, cv, is a statistical measure that determines the deviation of data from its mean and is defined as the ratio of standard deviation in a set of values to the mean of those values. We see in Figure 5.14 that for $\alpha > 0.95$, $cv < 0.03$, which means that standard deviation is 3% of the mean. In a standard normal distribution, 99.7% values lie within three standard deviations from mean. Therefore, almost all the values of the estimates lie within ±4.5% of the actual number of missing tags when $\alpha > 0.95$.

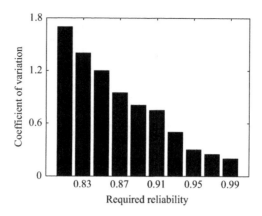

Figure 5.14   cv in estimates of missing tags

5.8.7   Comparison with tag ID collection protocol

RUN_D is faster than the fastest tag collection protocol, TH, whenever $T \geq 4$ for $\alpha > 0.9$ and $T \geq 7$ for $\alpha > 0.99$. This means that when the required reliability is 0.9, it is faster to execute TH only when the threshold T is less than 4. Similarly, when the required reliability is 0.99, it is faster to execute TH only when $T < 3$. This result is intuitive because when the threshold is small, RUN_D needs to execute more slots to ensure that it does not fail to detect a missing tag. Note that in most large scale settings, such as in large warehouses, containing tens of thousands of tags, the threshold T is set greater than 7 because it is quite likely that due to the communication errors between tags and readers, some tags may appear to be missing when they actually are not. Therefore, in large scale settings, RUN_D will always be faster than TH as long as $T \geq 7$. We achieved these results both theoretically and through simulations. Next, we first describe the theoretical relationship between the execution times (in terms of number of slots) of RUN_D and TH and then describe how we conducted the simulations to verify the theoretical results.

The authors of TH have shown that to collect the IDs of all tags in a population containing t tags, TH executes approximately $1.5t$ slots (refer to Figure 7(a) in [44]). Let k represent the number of slots TH executes to collect the IDs of $|\mathbb{U}| + |\mathbb{E}| - T$ tags. As per the results from [44], $k \approx 1.5 \times (|\mathbb{U}| + |\mathbb{E}| - T)$. To calculate the values of α and T for which the number of slots of RUN_D equal those of TH, we put $S_d = k$. As $S_d = f \times n$, thus, $f \times n = k$. Substituting $f = k/n$ and $p = 1$ in (5.8), we get an expression that is only a function of n. Solving this expression with (5.10), and putting $k = 1.5 \times (|\mathbb{U}| + |\mathbb{E}| - T)$, we get:

$$\alpha = 1 - 0.5^{1.5T \ln 2} \tag{5.29}$$

This equation expresses the relationship between the values of α and T for which the execution times of TH and RUN_D are equal. Note that this equation does not depend on the values of $|\mathbb{U}|$ and $|\mathbb{E}|$. Figure 5.15 shows a plot of this equation.

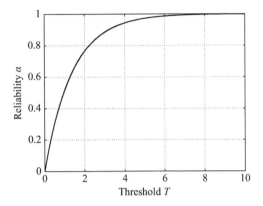

Figure 5.15 α vs. T when time of TH and RUN equal

To validate this equation, we executed both RUN_D and TH by varying $|\mathbb{E}|$ from 500 to 10,000, $|\mathbb{U}|$ from 0 to 10,000, and T from 100 to $|\mathbb{E}|/2$. For each set of values of $|\mathbb{E}|$, $|\mathbb{U}|$, and T, we executed both RUN_D and TH 500 times to get their corresponding average number of slots. In these simulations, to calculate the optimal values of n and f for RUN_D, for each set of values of $|\mathbb{E}|$, $|\mathbb{U}|$, and T, we put the value of T in (5.29) to obtain the corresponding value of α. We observed from these simulations that the execution times of RUN_D and TH for each set of values of $|\mathbb{E}|$, $|\mathbb{U}|$, and T, were almost the same. The maximum difference between the average values of the number of slots for RUN_D and of TH for any set of values of $|\mathbb{E}|$, $|\mathbb{U}|$, and T was not more than 0.01%. This validates (5.29) and consequently the observation mentioned at the start of this subsection.

5.9 Conclusions

The key technical contribution of this chapter is in proposing a protocol to detect and identify missing tag events in the presence of unexpected tags. This chapter represents the first effort on addressing this important and practical problem. The key technical depth of this chapter is in the mathematical development of the theory that RUN_D and RUN_I are based upon. The solid theoretical underpinning ensures that the actual reliability of RUN_D is greater than or equal to the required reliability. We have proposed a technique that our protocol uses to handle large frame sizes to ensure compliance with the C1G2 standard. We have also proposed a method to implicitly estimate the size of the unexpected tag population without requiring an explicit estimation phase. We implemented RUN_D and RUN_I and conducted side-by-side comparisons with four major prior missing tag detection and identification protocols even though none of the existing protocols handle the presence of unexpected tags. Our protocols significantly outperform all prior protocols in terms of actual reliability and detection time.

Chapter 6

RFID detection—unknown tags

6.1 Introduction

6.1.1 Background

Radio-frequency identification (RFID) technology is a type of wireless technique that can automatically identify or track the RFID tags attached to objects or even humans. Compared with the conventional bar-code systems, RFID technique owns many attractive advantages, such as remote and multiple access, simple computational ability, and nonsight limitation, etc. Therefore, it is widely used in the localization [60,101,102], warehouse monitoring [67,103,104] and supply chain management [105–108]. A typical RFID system consists of a back-end server, a single (or multiple) reader(s), and a large number of tags [70]. The tags are attached to items, and each one has a small memory to store its unique ID (96-bit) for individual identification and some information (e.g., product price, shelf-life, etc.) of the tagged item. The RFID tags are generally classified into two types: active tags and passive tags. The active RFID tag uses an internal battery to continuously power the RF communication circuitry, whereas the passive RFID tag relies on RF energy from the reader to power its circuitry.

6.1.2 Motivation and problem statement

In most RFID-enabled applications such as inventory control and warehouse management, normally, the ID information of all tags is stored in a database of the back-end server. When some tags are moved out, the corresponding ID information are deleted from the database; when new tags are moved in, the corresponding ID information obtained from tag identification protocols [78,109] are added into database. However, due to errors in the communication link, e.g., caused by the obstacles in the radio path, some new tags may be left unread [110,111], and thus unknown tags appear. On the other hand, the misplacement of tagged items can also result in the unknown tags because the misplaced tags are "strangers" to the reader locating in this region. The chilled food will quickly decay if it is misplaced at the zone with no refrigeration equipment and not discovered in time. More seriously, in the warehouse management, a tagged lighter that is misplaced in the area of alcohol may cause fatal safety problem [112]. For reader, both the unread tags and the misplaced tags are *unknown tags*.

The unknown tag problem (*UTP*) can be generally classified into two subproblems: detection and identification. The unknown tag detection problem concentrates on discovering the existence of the unknown tags. In contrary, the unknown tag identification problem aims at pinpointing the exact unknown tags. Intuitively, the identification process consumes more time (and energy when active tags are used) than detection process, because identification needs to report all information about unknown tags, while detection only needs to report the existence of unknown tags (i.e., *YES* or *NO*). The heavy identification process sometimes does not discover any unknown tags and thus wastes a lot of time because there are no unknown tag at all. Inspired by [83], a wise way is to periodically perform the tiny unknown detection protocol, and the identification protocol (e.g., BUIP-CF) is invoked only when the detection protocol reports the existence of unknown tags.

In this chapter, we study the problem of unknown tag detection, which is formally defined as follows. *We target at providing a detection protocol to discover the existence of unknown tags with a reliability of at least β if λ or more unknown tags appear in the RFID system that consists of N known tags.*

6.1.3 Existing work and limitations

The Single Echo-based Batch Authentication (SEBA) protocol proposed in [113] uses the well-known framed-slotted Aloha communication mechanism to detect the existence of unknown tags (called counterfeit tags in [113]). The unexpected response received by the reader is treated as the evidence of unknown tags. SEBA+ [114] exploits the standard bloom filter (BF) technique to generalize the SEBA. In SEBA+, each tag randomly selects $r \geq 1$ slots to respond within a frame. As long as one responding location of an unknown tag is not covered by response(s) from known tags (called legitimate tags in [113]), the existence of unknown tags is discovered. SEBA is a special case ($r = 1$) of SEBA+.

These two protocols have three drawbacks: (1) The BF is constructed distributively by tags on the *slot level*. According to Philips I-Code specification [79], any two consecutive transmissions (from a tag to a reader or vice versa) are separated by a waiting time $t_0 = 302$ μs. Hence, constructing $BF_{T \Rightarrow R}$ (from tags to reader) is time consuming. For example, if we want to construct a BF with 1,000 bits, the time cost of $BF_{T \Rightarrow R}$ is $1,000 \times (t_0 + t_{bit})$, where t_{bit} is the time of transmitting 1-bit information from a tag to the reader. It involves a large amount of waiting time, which contributes to the low time efficiency. (2) SEBA and SEBA+ leverage the expected empty slots to detect the unknown tags. To guarantee the required detection accuracy, the ratio of expected empty slots to the whole frame size should be kept at a certain level. Therefore, the frame size is necessary to be proportional to the tag population size, which leads to the poor scalability. (3) Energy efficiency, an important metric when active tags are used, is not considered in [79,113].

6.1.4 Main contributions

According to the investigation of the prior work, we think the problem of unknown tag detection is of practical importance and still soliciting new efficient solutions.

To this end, this chapter studies the problem of unknown tag detection and makes the following contributions:

1. We combine the standard BF and sampling idea to propose the sampling BF (SBF), in which no element is necessary to use k times hashing; however, each of the k hashings depends on the sampling probability.
2. We exploit the new filtering technique to propose the SBF-based unknown tag detection protocol (SBF-UDP), in which the reader constructs a filter according to the IDs of known tags. Then, the filter is broadcasted to all tags for verifying their membership.
3. This chapter takes both time efficiency and energy efficiency into account. Sufficient theoretically analysis is proposed to investigate the parameter settings to minimize the time cost (or energy consumption) while guaranteeing the predefined detection reliability.
4. Extensive simulations are conducted to evaluate the performance of the proposed protocol. The experimental results demonstrate that the proposed protocol considerably outperforms the existing related protocols in terms of both time efficiency and energy efficiency.

The rest of this chapter is organized as follows. The related work is reviewed in Section 6.2. Section 6.3 describes the preliminary knowledge including system model and the used performance metrics. We propose the SBF-UDP as well as the theoretical analyses of the parameter settings in Section 6.4. In Section 6.5, extensive simulation experiments are conducted to evaluate the performance of the proposed SBF-UDP. This chapter is concluded in Section 6.6.

6.2 Related work

RFID is an emerging technology that is widely used in many monitoring applications. There are two basic monitoring tasks that solicit efficient solutions: missing tag problem and UTP.

Missing-tag detection. In recent years, many efforts have been made to address the missing-tag problem. Tan *et al.* proposed the trust reader protocol (TRP) to detect the missing-tag event with a predefined probability β when the number of the missing tags exceeds a given threshold [115], which also inspired the problem formulation of this chapter. To improve the time efficiency and energy efficiency of TRP, Luo *et al.* introduced the sampling idea, and thus proposed the efficient missing-tag detection (EMD) protocol, where they used the detection result on the sampled tags to probabilistically reflect the whole intactness [116]. However, EMD still has a large room to be improved because it contains a large proportion of expected empty/collision slots that cannot be used in missing-tag detection. To overcome this deficiency, Luo *et al.* studied a multihashing-seed approach to reduce the useless empty slots and collision slots involved in the EMD protocol and thus proposed the multiseed missing-tag detection (MSMD) protocol [83]. These protocols (TRP [115], EMD [116], and

MSMD [83]) concentrate on discovering the missing-tag event, instead of pinpointing which tags are missing.

Missing-tag identification. The iterative ID-free protocol proposed in [94] is a variant of the framed-slotted Aloha protocol and is able to pinpoint the exact missing tags. The protocols proposed by Zhang *et al.* in [98] accelerate the protocol's execution by leveraging the collaboration of multiple readers. To improve the slot utilization, an efficient collision reconciling method was proposed to change some collision slots into singleton slots that can be used for the identification of missing tags [117]. Zheng *et al.* investigated the compressive sensing technique to perform the identification at the physical layer [99].

Unknown tag identification. The UTP is also very important in practice [118]. As aforementioned, the unknown tagged objects such as misplaced or left unread items could cause serious economic loss or security issues. The existing protocols dedicated for unknown tag detection have been reviewed in Section 6.1. In terms of unknown tag identification, Sheng *et al.* exploited the framed-slotted Aloha mechanism to propose a CU (Collect Unknown tags) [119]. The tags responding the expected empty slots are necessary to be unknown tags and will keep active to be collected by enhanced dynamic framed-slotted Aloha (EDFSA) protocol [77]. Multiple rounds are required to achieve desired identification accuracy. In the CU protocol, only the expected empty slots are used to detect the unknown tags. However, other slots are not fully explored. Moreover, all known tags participate in each round, which interferes the unknown tag detection. To overcome these drawbacks of the CU protocol, the BUIP-CF protocol proposed in [120] uses, in addition to the expected empty slots to label the unknown tags, also the expected singleton slots to deactivate the known tags (preventing them from interfering the detection of unknown tags). Specifically, if one and only one tag responds in an expected singleton slot, this must be known tag. Then, the reader sends an acknowledgement (ACK) signal to deactivate it (i.e., to enter the sleep state). However, if one or more tags respond in an expected empty slot, all of them must be unknown tags. Then, the reader sends a negative acknowledgement (NACK) signal to label them (i.e., telling them not to participate in the next round, but still remain active). In a round, some known/unknown tags could be deactivated/labeled, and they will not participate in the following rounds; the other tags will participate in the next round. This process is repeated for multiple rounds until all known tags are deactivated. In contrary, all the unknown tags keep active and will be collected by the EDFSA protocol [77]. Liu *et al.* proposed to use bit vector (from reader to tags) to fast filter out the unknown tags [121] for the purpose of unknown tag identification. Further, they investigated a bitwise XOR filter technique, the "1s" in the XOR filter cannot only label the unknown tags but also accelerate the unknown ID collection [118].

Bloom filter technique. A BF [122–124] is a well-known data structure that is very popular in database applications and also receives widespread attention in the networking literature. BF probabilistically represents a set of n elements $Y = \{y_1, y_2, \ldots, y_n\}$, which can be used to test set membership. Specifically, the BF compresses this set into a filter vector with w bits by hashing each element in Y into the vector using k hashing functions $H_1, H_2, \ldots, H_k$. As illustrated in Figure 6.1(a), a bit in the vector is set to "1" if at least one element is hashed to that index in the vector;

otherwise, it is set to "0." When checking whether a given element y belongs to the set Y, we compute $H_1(y), H_2(y), \ldots, H_k(y)$ and assert $y \in S$ if and only if all these k bits are "1s" in the vector; otherwise, $y \notin Y$. BF technique is relatively lightweight and has potential to be used on RFID devices. For example, [103,106] use BF or its variant to search the exact wanted tags in a given RFID system.

6.3 Preliminary

6.3.1 System model and assumption

We assume a large-scale RFID system that consists of a back-end server, a reader, N known tags, and M unknown tags, where usually $M \ll N$. Let T_Δ denote the known tag set, i.e., $T_\Delta = \{t_1, t_2, \ldots, t_i, \ldots, t_N\}$, and the number N as well as ID information in T_Δ is available in a database on the back-end server. The unknown tag set is denoted as T_Λ, i.e., $T_\Lambda = \{tu_1, tu_2, \ldots, tu_i, \ldots, tu_M\}$, whereas both the number M and the specific ID information in T_Λ are not known in advance. Each tag has a unique ID and is equipped with the same *uniform* hash generator $H(\cdot)$. We assume the reader has adequate power to interrogate all the tags including the known ones and the unknown ones. Moreover, the reader communicates with the back-end server via a high-rate network link and has access to the ID information of all known tags.

The reader communicates with the tags in a time-slotted way, where the slots are synchronized by the periodical *"end slot"* commands broadcasted by the reader. According to the specification of the Philips I-Code system [79], the wireless transmission rate from a tag to a reader is 53 kb/s, and the rate from a reader to a tag is 26.5 kb/s. Any two consecutive transmissions (from a tag to a reader or vice versa) are separated by a waiting time t_0 of 302 μs [112]. The duration of a slot that supports transmission of μ-bit data is $t_0 + \mu t_{\text{bit}}$, where t_{bit} is the time for transmitting 1 bit. Usually, μ is less than 96, i.e., the length of tag ID. Following prior literature [94], the slots are classified into *tag slots, long-response slots,* and *short-response slots*. The length of a tag slot is denoted as t_{tag}, which allows the transmission of a tag ID (96 bits), either from the reader to the tags or from a tag to the reader. The length of a long-response slot is denoted as t_{long} and is sufficient to transmit 10 bits information. The length of a short-response slot is denoted as t_{short} and allows the transmission of only 1 bit information. In this chapter, t_{tag} is set to be 4 ms for transmission of a tag ID (96 bits) from a tag to a reader or vice versa. Similarly, t_{long} and t_{short} are set to be 0.7 and 0.4 ms, respectively.

6.3.2 Energy consumption model

Because the battery of a reader can be easily recharged or the reader may even use an external power source [83], the energy consumed by the reader is ignored in this chapter. Accordingly, we only consider the energy consumption of the battery-powered active tags, particularly, the known tags. During a slot, an active tag has two types of states: *awake* state and *sleep* state [125]. Specifically, a tag needs to be in the awake state (i.e., its CPU works at full energy and the radio remains active) for communication. An awake tag may operate one of the three actions during a certain slot: transmitting data to the reader; receiving data from the reader; or just listening the channel for the periodical "end-slot" commands broadcasted by the reader. Since the radio scanning consumes most of the energy, the above actions of an awake tag almost consume the same amount of energy. Similar assumptions are also made in [126]. Let ω denote the energy consumption of an awake tag during a tag slot. Since the length of a short-response slot is $\frac{1}{10}$ of the tag slot, the energy consumption of an awake tag during the short-response slot is about $\frac{1}{10}\omega$. For similar reason, the energy consumed in a long-response slot is about $\frac{7}{40}\omega$. To conserve battery power, the tag can enter the sleep state, where the CPU works in a low power mode and radio reception is disabled. The ratio of energy consumed between the awake and sleep states is typically on the order of 100 or more [125], so we neglect the energy consumption of an asleep tag. We use E to denote the energy consumption of N known tags, which is given as $E = \sum_{i=1}^{N} [\eta_{i1} \cdot (\frac{1}{10}\omega) + \eta_{i2} \cdot (\frac{7}{40}\omega) + \eta_{i3} \cdot \omega]$, where η_{i1}, η_{i2}, and η_{i3} indicate the number of the short-response slots, the long-response slots, and the tag slots that tag t_i keeps awake for, respectively. The used notations are summarized in Table 6.1.

6.3.3   Performance metrics

We consider two important performance metrics: (1) *execution time*: we obviously desire to discover the unknown tag event as soon as possible, and thus we are able to take timely countermeasures, e.g., replacing the chilled food to the zone equipped with freezers. (2) *Energy consumption*: the battery volume of active tag is typically limited, and replacing batteries for a large number of tags is quite laborious. Therefore,

Table 6.1   Notations used in this chapter

Symbols	Descriptions
ID	ID information of a RFID tag
N	Number of known tags in the system
M	Number of unknown tags in the system
λ	Tolerance threshold
β	Detection reliability
T_Δ	Set of all known tags in the system
T_Λ	Set of all unknown tags in the system
t_{tag}	Length of a tag slot
t_{long}	Length of a long-response slot
t_{short}	Length of a short-response slot
ω	Energy consumption of an awake tag during a tag slot
$H(\cdot)$	Hash generator that follows a uniform random distribution
X	A large constant preconfigured in the RFID tags
k	Number of employed hashing functions
p	Sampling probability
$\langle p_t, k_t \rangle$	Optimal pair that minimizes the time cost of SBF-UDP
$\langle p_e, k_e \rangle$	Optimal pair that minimizes the energy cost of SBF-UDP

it is desirable to reduce the energy consumption as much as possible. The protocols proposed in this chapter are not limited to the active tags. If only passive tags are used in an application, the time efficiency becomes the only performance metric.

6.4 A sampling bloom filter-based unknown tag detection protocol

In this section, we first propose the *SBF* which generalizes the standard BF. Then, we present the detailed design of the *SBF-UDP*. After that, we theoretically investigate how to configure the system parameters, e.g., the number of hashing functions, the sampling probability, the filter length, to guarantee the predefined detection reliability. Finally, we analyze the time cost and energy cost of our SBF-UDP, respectively.

6.4.1 Overview of the sampling bloom filter

Given a set $Y = \{y_1, y_2, \ldots, y_n\}$ containing n different elements, the principle of the SBF is presented as follows. As illustrated in Figure 6.1(b), each element is mapped to the filter using k hashing functions. However, different from the *standard* BF, when "inserting" a certain element into the filter, each of the k mappings is really implemented with a sampling probability p. For example, the element y_1 is hashed to the fifth bit using $H_1(\cdot)$, but this mapping is not sampled, then the fifth bit will not be set to "1." When checking whether a given element y belongs to Y, as long as one of the k mappings is sampled and the corresponding bit is "0," we can assert that $y \notin Y$; otherwise, $y \in Y$. Clearly, the proposed SBF is a general version of BF. In other words, the standard BF is a special case ($p = 1$) of the proposed SBF.

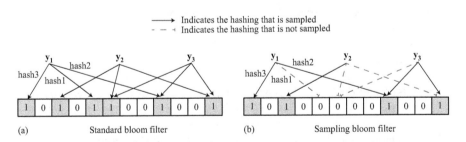

Figure 6.1 *Standard bloom filter vs. sampling bloom filter. (a) Standard bloom filter: each element is pseudorandomly compressed into the filer using k different hashing functions; (b) sampling bloom filter: each element is also pseudorandomly mapped to the filter with k hashing functions; however, each mapping is implemented with a sampling probability p, and the sampling process is also pseudorandom*

6.4.2 Protocol design of SBF-UDP

The proposed SBF-UDP consists of three stages: (1) *constructing the SBF*; (2) *verifying the tag identity*; and (3) *announcing unknown identity*.

6.4.2.1 Constructing the sampling bloom filter

The reader has access to a database that stores the ID information of all known tags. The reader then compresses each known tag ID to the f-bit filter F using k uniform hashing functions. Specifically, an arbitrary known tag t_i with ID_i is mapped to the locations $l_{i1}, l_{i2}, \ldots, l_{iu}, \ldots, l_{ik}$, where $l_{iu} = H_u(ID_i, R_1) \mod f$, $u \in [1, k]$, and R_1 is a random seed. The k bits are called the *representative* bits of tag t_i. Note that, no representative bit is necessary to be set to "1," which actually depends on the sampling result. Given a sampling probability $p \in (0, 1]$, the reader calculates an integer $x = \lceil X \times p \rceil$, where X is a large constant that is preconfigured in the tag chip [116]. The representative bit with location l_{iu} is sampled and set to "1" only when $H_u(ID_i, R_2) \mod X \leq x$, where R_2 is another random number used for sampling process; otherwise, the reader does not change this bit. The SBF is completed after all the known IDs are *inserted* into the filter according to the method described above.

6.4.2.2 Verifying the tag identity

Then, the reader broadcasts the employed parameters $\langle R_1, R_2, k, f, x \rangle$ along with the constructed f-bit filter to all RFID tags. Each tag uses the same k hash functions $H_1(\cdot) \sim H_k(\cdot)$ and the received parameters to check whether it belongs to the known tag set. Specifically, for an arbitrary tag, it first calculates $H_u(ID, R_2) \mod X$, $u \in [1, k]$. If the hashing result is equal to or less than x, it means the uth representative bit of this tag is sampled. Then, the tag further calculates $l_u = H_u(ID, R_1) \mod f$. If the l_u^{th} bit is "0," this tag will be aware of its unknown identity; otherwise, it will check its other representative bits using the same method. As long as one of the representative bits is sampled but turns out to be "0," the tag is necessary to be unknown tag.

6.4.2.3 Announcing unknown identity

At the end of this protocol, the unknown tags that are aware of their unknown identity will respond to the reader to announce its unknown identity. If the reader senses a busy slot carrying announcement(s), it will successfully detect the existence of unknown tag(s).

Clearly, the actual number of unknown tags has significant impact on the detection results of our SBF-UTD. The more unknown tags there are in the system, the more likely our SBF-UTD detects the existence of unknown tags. (1) If there is *no* unknown tag, SBF-UTD will not report the existence of unknown tags. (2) If the number of unknown tags is less than the threshold λ, SBF-UTD can still have a chance to discover the existence of unknown tags. But we do not guarantee the required reliability of β. (3) If the number of unknown tags is larger than the threshold λ, theoretical analysis in the next section can guarantee that the actual reliability of SBF-UTD is larger than β. No matter how many unknown tags appear in the system, the processes of SBF-UTD is always the same because unknown tags announce their unknown identity at the end of our protocol. Even if there is *no* unknown tag, SBF-UTD still needs to go through the above three stages. The execution time is independent of the number of unknown tags.

6.4.3 Investigating the detection accuracy

There are two types of false detection for the unknown tag-detection problem. (1) The detection protocol reports the existence of unknown tags, although there is no unknown tag (i.e., "false alarm"). (2) The detection protocol does not discover any unknown tags, despite the presence of unknown tags (i.e., "miss-detection"). SBF-UTD does not incur *false alarms*, because the sampled representative bits of a known tag will always be "1s." On the other hand, our protocol indeed suffers from the *miss-detection*, which occurs when all the representative bits of unknown tags are "occupied" by known tags. We formally analyze the miss-detection in this section.

We use $P(N,M,f,k,p)$ to denote the probability of successfully discovering the existence of unknown tag(s) when there are N known tags and M unknown tags, the filter length is f, the sampling probability is p, and the k uniform hash functions are used. In the following, we theoretically propose the expression of $P(N,M,f,k,p)$. Let P_{false} denote the probability that an arbitrary unknown tag eventually passes the detection. An unknown tag can pass the detection only when *each of its k* representative bits satisfies one of the following conditions: (1) not sampled; (2) if sampled, at least one known tag is hashed to the same location, which sets the representative bit to "1." P_{false} is expressed as follows:

$$P_{\text{false}} = \left\{ (1-p) + p\left[1 - \left((1-p) + p\left(1 - \frac{1}{f}\right)\right)^{Nk}\right] \right\}^k$$

$$= \left(1 - pe^{-\frac{Nkp}{f}}\right)^k .$$

(6.1)

As long as one of the M unknown tag is discovered, the existence of unknown tags will be successfully reported. Hence, the detection probability $P(N,M,f,k,p)$ can be given as

$$P(N,M,f,k,p) = 1 - P_{\text{false}}^M = 1 - (1 - pe^{-\frac{Nkp}{f}})^{Mk}$$

(6.2)

According to (6.2), we find that $P(N,M,f,k,p)$ monotonically increases as M increases. Hence, when the number M of unknown tags exceeds the threshold λ, we have $P(N,M,f,k,p) \geq P(N,\lambda,f,k,p)$. Obviously, if we could guarantee $P(N,\lambda,f,k,p) \geq \beta$, the actual detection probability will be larger than β for any $M \geq \lambda$. By solving the inequality $P(N,\lambda,f,k,p) \geq \alpha$, we have

$$f \geq \frac{-Nkp}{\ln\left[\frac{1-(1-\beta)^{\frac{1}{\lambda k}}}{p}\right]}$$

(6.3)

6.4.4 Analyzing the performance of SBF-UDP

In this subsection, we analyze the time cost and the energy cost of the proposed SBF-UDP, respectively. Moreover, we investigate how to minimize them.

6.4.4.1 Time cost

Following [113,114,119,120], we only consider the time consumed by the wireless communication between the reader and the tags and exclude the time consumed by computation. In the stage of *constructing the SBF*, one tag slot t_{tag} is adequate for the reader to broadcast the initialization parameters. The f-bit filter will be divided into S segments of 96 bits to be sequentially transmitted in S tag slots [94] when its length is larger than 96 bits (the maximum number of bits that can be transmitted in a tag slot), where $S = \lceil \frac{f}{96} \rceil$. Then, the time for by transmitting the filter is $\lceil \frac{f}{96} \rceil \times t_{\text{tag}}$, where t_{tag} can afford transmitting 96-bit data.

In the stage of *verifying the tag identity*, the reader waits one short-response slot t_{short} for listening the expected announcement from the unknown tags. Hence, the time cost T of SBF-UDP is given as follows:

$$T = t_{\text{tag}} + \left\lceil \frac{f}{96} \right\rceil \times t_{\text{tag}} + t_{\text{short}}. \tag{6.4}$$

6.4.4.2 The minimum time cost

According to (6.3), f should be set larger than $-Nkp / \left\{ \ln \left[\frac{1-(1-\beta)^{\frac{1}{\lambda k}}}{p} \right] \right\}$ so as to meet the predefined accuracy. Equation (6.4) shows that T is an increasing function with respect to f. Hence, given N, λ, β, k, and p, the filter length f should be minimized to $-Nkp / \left\{ \ln \left[\frac{1-(1-\beta)^{\frac{1}{\lambda k}}}{p} \right] \right\}$ to achieve the minimum time cost $T_{\min}$. Note that, this is just a *local* optimization with fixed p and k. As illustrated in Figure 6.2, different

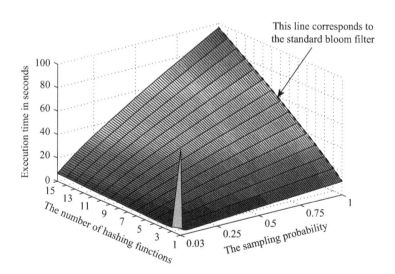

Figure 6.2   Investigating the impact of p and k on the time cost of SBF-UDP, where $N = 10,000$, $\lambda = 100$, $\beta = 95\%$, p varies from 3% to 100%, and k changes from 1 to 15

parameter pairs $\langle p, k \rangle$ result in different time cost of SBF-UDP. Note that, the dotted line in Figure 6.2 actually corresponds to the use of the standard BF (i.e., $p = 1$), which reveals the superiority of our SBF. Then, the next key task is to search the optimal pair $\langle p_t, k_t \rangle$ to get the *overall* minimum time cost of SBF-UDP. Note that, if the sampling probability p is too small, the detection accuracy cannot be larger than the desired reliability β for any values of k and f. Such a small sampling probability p cannot be used. In what follows, we derive the *lower bound* p_{min} of the sampling probability.

$$f \geq \frac{-Nkp}{\ln\left[\frac{1-(1-\beta)^{\frac{1}{\lambda k}}}{p}\right]} > 0 \tag{6.5}$$

$$\Rightarrow p > 1 - (1 - \beta)^{\frac{1}{\lambda k}}.$$

Hence, the lower bound p_{min} on the sampling probability is $1 - (1 - \beta)^{\frac{1}{\lambda k}}$. For instance, if the tolerance number λ is set to 20, the number k of used hashing functions is 5, and the detection reliability β is set to 95%, the minimum sampling probability p_{min} is about 3%.

6.4.4.3 Energy cost

Recall that the long filter is divided into S segments of 96 bits to be sequentially transmitted, where $S = \lceil \frac{f}{96} \rceil$. Each tag remains awake before successfully receiving all the segments containing its representative bits. For an arbitrary known tag (we do not consider the energy consumption of the unknown tags), let L be the index of the *last* segment that contains its representative bit(s). Clearly, this tag has to keep active for L tag slots to receive the first L segments before entering the sleep state. The goal is to calculate the expectation of L. This can be treated as a problem of allocating k balls into S bins. The bins are numbered from 1 to S. Each of the k balls is thrown with probability p into one of the S bins following uniform distribution. Note that, one bin can contain more than one balls. **Question:** *Which bin is the last one that contains ball(s) on average?* We propose Theorem 6.1 to answer the question.

Theorem 6.1. Assuming there are k balls to be thrown into S bins numbered from 1 to S. For a certain ball, it is determined to throw with a probability p. If we determine to throw it, the S bins have the same chance $\frac{1}{S}$ to obtain this ball. After tackling all the balls, the index of the last bin containing ball(s) is denoted as L. The expectation $E(L)$ of L is $S - \sum_{i=0}^{S-1} (1 - p + \frac{ip}{S})^k$.

Proof. Let $P[L = u]$ be the probability that the uth bin is the last bin containing at least one ball, where $u \in [0, S]$. $u = 0$ means no balls are thrown. $P[L = u]$ is given as follows:

$$P[L = u] = \begin{cases} \sum_{i=1}^{k} \binom{k}{i} \left(p\frac{1}{S}\right)^i \left[(1-p) + p\frac{u-1}{S}\right]^{k-i} & u \in [2, S] \\ \sum_{i=1}^{k} \binom{k}{i} \left(p\frac{1}{S}\right)^i (1-p)^{k-i} & u = 1 \\ (1-p)^k & u = 0 \end{cases} \tag{6.6}$$

According to (6.6), $P[L = u]$ can be simplified to:

$$P[L = u] = \begin{cases} \left(1 - p + \frac{up}{S}\right)^k - \left[1 - p + \frac{p(u-1)}{S}\right]^k & u \in [1, S] \\ (1 - p)^k & u = 0 \end{cases}$$

(6.7)

The expectation $E(L)$ is given as:

$$E(L) = \sum_{u=0}^{S} u \times P[L = u] = S - \sum_{i=0}^{S-1} \left(1 - p + \frac{ip}{S}\right)^k.$$

(6.8)

□

Corollary 6.1. For fixed k and p, $E(L)$ is an increasing function with respect to S.

Proof. Given a fixed k, $E(L)$ becomes a function with respect to S, which is denoted as $Q(S)$. For an arbitrary positive integer S, if $Q(S + 1) > Q(S)$ always holds, $E(L)$ is proved to be an increasing function with respect to S. According to (6.8), $Q(S + 1) - Q(S)$ is given as:

$$\begin{aligned} &Q(S + 1) - Q(S) \\ &= \left[(S + 1) - \sum_{i=0}^{S} \left(1 - p + \frac{ip}{S+1}\right)^k\right] - \left[S - \sum_{i=0}^{S-1} \left(1 - p + \frac{ip}{S}\right)^k\right] \\ &= \underbrace{1 - \left(1 - \frac{p}{S+1}\right)^S}_{>0} - \underbrace{\sum_{i=0}^{S-1} \left[\left(\frac{ip}{S+1}\right)^k - \left(\frac{ip}{S}\right)^k\right]}_{<0} > 0. \end{aligned}$$

(6.9)

Therefore, $Q(S)$ (i.e., $E(L)$) is an increasing function with respect to S. □

Recall the problem of allocating k balls to S bins. The last bin containing ball(s) is expected to be the $E(L)$th bin, determined by (6.8). Known tags have to keep awake before receiving the last segment ($E(L)$th segment) containing its representative bits. Overall, they have to remain awake for *one* tag slot to receive the initialized parameters and $E(L) = S - \sum_{i=0}^{S-1} \left(1 - p + \frac{ip}{S}\right)^k$ (*expectation value*) more tag slots before entering the sleep state. The expected energy consumption of a known tag during one execution of SBF-UDP is $[1 + E(L)] \times \omega$. For N known tags in total, the energy cost E of SBF-UDP (excluding the energy consumption of unknown tags) is:

$$\begin{aligned} E &= N \times [1 + E(L)] \times \omega \\ &= N \times \left[1 + S - \sum_{i=0}^{S-1} \left(1 - p + \frac{ip}{S}\right)^k\right] \times \omega. \end{aligned}$$

(6.10)

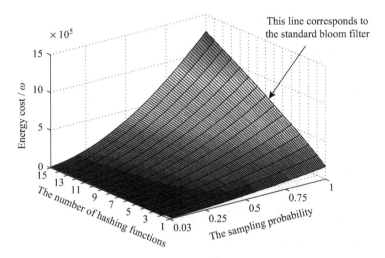

Figure 6.3 Investigating the impact of p and k on the energy cost of SBF-UDP, where N = 10,000, λ = 100, β = 95%, p varies from 3% to 100%, and k changes from 1 to 15

6.4.4.4 The minimum energy cost

When the battery-powered active tags are used, as aforementioned, we may need to minimize the energy cost, thereby prolonging the lifetime of the RFID system. Obviously, the energy cost E of SBF-UDP is an increasing function with respect to $E(L)$. Moreover, according to Corollary 6.1, $E(L)$ is also an increasing function with respect to S. Hence, the energy cost E is an increasing function with respect to S, where $S = \lceil \frac{f}{96} \rceil$. Similar to the analysis of the minimum time cost T in Section 6.4.4.2, the filter length f should be minimized to $-Nkp/\{\ln[\frac{1-(1-\beta)^{\frac{1}{\lambda k}}}{p}]\}$, thereby achieving the minimum energy cost while meeting the predefined detection accuracy. Note that, this is also a *local* optimization with given k and p. As illustrated in Figure 6.3, the parameters $\langle p, k \rangle$ significantly affect the energy cost of SBF-UTP. Again, to get the *overall* minimum energy cost E, searching the optimal parameter pair p_e and k_e is necessary.

In this chapter, we propose a simple Algorithm 1 to find the optimal parameter pair $\langle p_t, k_t \rangle$ to achieve the overall minimum execution time and $\langle p_e, k_e \rangle$ to achieve the overall minimum energy cost. Specifically, steps 1–4 in Algorithm 1 initialize variables. δ specifies the maximum deviation between the calculated p and its real optimal value. k is typically less than 15 [127]. Therefore, we traverse all values of k between 1 and 15 in step 5. For each k, steps 6–23 in Algorithm 1 test p values at δ intervals apart to find the minimum time cost as well as energy cost, meanwhile recording the corresponding p and k. Specifically, for fixed values of p and k, we calculate the corresponding time cost and energy cost from steps 9 to 12. During steps 13–17, we update the $T_{\min}$ and record the corresponding p_t and k_t. During steps 18–22, we update the $E_{\min}$ and record the corresponding p_e and k_e.

Algorithm 1: Find the optimal sampling probability p and hashing count k of SBF-UDP.

Input: The number N of known tags; tolerance number λ; the reliability β.
Output: The optimal $\langle p_t, k_t \rangle$ that minimizes the time cost; The optimal $\langle p_e, k_e \rangle$ that minimizes the energy cost.

1: $T_{\min} = E_{\min} = +\infty$;
2: $k_t = k_e = -1$;
3: $p_t = p_e = -1$;
4: $\delta = 0.001$;
5: **for** each $k \in [1, 15]$ **do**
6:     $p_{\min} = 1 - (1 - \beta)^{\frac{1}{\lambda k}}$;
7:     $p = 1$;
8:     **while** $p > p_{\min}$ **do**
9:         $f = -Nkp / \{\ln \left[\frac{1 - (1-\beta)^{\frac{1}{\lambda k}}}{p} \right] \}$;
10:        $S = \lceil \frac{f}{96} \rceil$;
11:        $T = t_{\text{tag}} + S t_{\text{tag}} + t_{\text{short}}$;
12:        $E = N \times [1 + S - \sum_{i=0}^{S-1} (1 - p + \frac{ip}{S})^k] \times \omega$;
13:        **if** $T < T_{\min}$ **then**
14:            $T_{\min} = T$;
15:            $k_t = k$;
16:            $p_t = p$;
17:        **end if**
18:        **if** $E < E_{\min}$ **then**
19:            $E_{\min} = E$;
20:            $k_e = k$;
21:            $p_e = p$;
22:        **end if**
23:     **end while**
24: **end for**
25: return $\langle p_t, k_t \rangle$ and $\langle p_e, k_e \rangle$;

The computational complexity of Algorithm 1 is $O(\frac{1}{\delta})$. The computer does not need to frequently run Algorithm 1 (offline) because the parameters N, λ, and β are not changed often in practice for monitoring applications.

6.4.4.5 Joint optimization of time and energy costs

Joint consideration of the time efficiency and energy efficiency is justified when the battery-powered active tags are used. For example, we desire to find optimal $\langle k, p \rangle$ to minimize energy while keeping the detection time within a given *upper threshold*. The joint optimization algorithm can be designed by slightly modifying Algorithm 1. Specifically, we only need to replace the condition at line 13 in Algorithm 1 by a new condition "**if** $T < T_{\min}$ and $E < E_{\text{threshold}}$ **then**," where $E_{\text{threshold}}$ is the maximum energy cost that is allowed. Similarly, we also replace the condition at line 18 in Algorithm 1 by a new condition "**if** $E < E_{\min}$ and $T < T_{\text{threshold}}$ **then**," where $T_{\text{threshold}}$ is the maximum time cost that is allowed.

6.5 Performance evaluation

Extensive simulation experiments are conducted to evaluate the performance of the proposed SBF-UDP in this section. We simulated the experimental conditions following related literature [113,114,119,120]: (1) considering a single reader in the simulations and assuming it has adequate power to interrogate with all RFID tags; (2) the signal interference between the adjacent RFID tags is ignored. In the following subsections, we first consider an error-free communication channel between the reader and tags. Numerical results are provided to show the advantages of our SBF over standard BF. Then, we conduct simulations to compare the proposed SBF-UDP with prior schemes [113,114,119,120] in terms of time efficiency and energy efficiency. Note that, because the identification-based protocols are far from efficient, we do not compare the proposed SBF-UDP with them. After that, experiments are conducted to show that the proposed SBF-UDP indeed achieves the required detection accuracy. Finally, we conduct simulations taking the nonperfect communication channel into consideration and evaluate the impact of channel error on our SBF-UTD protocol. A simple but effective countermeasure is proposed to mitigate the negative impact of channel error, and its effectiveness is also evaluated via simulations.

6.5.1 Demonstrating the advantages of sampling bloom filter

In this set of simulations, we provide numerical results to show the advantages of our SBF over conventional standard BF. As illustrated in Figure 6.4, we vary the sampling probability p from $p_{\min}$ to 1. For the purpose of clarity, we only vary the hash number k from 1 to 4. The numerical results in Figure 6.4 reveal that both time cost and energy cost are *convex* function with respect to p. The time (or energy) cost of traditional standard BF corresponds to $p = 1$. A too small sampling probability p will lose too much information, and thus consuming more time and energy. A proper setting of p will make our SBF perform much better than the standard BF.

6.5.2 Comparing with the prior related protocols

Besides SEBA [113] and SEBA+ [114], we also compare our detection protocol with two representative identification protocols, i.e., CU [119] and BUIP-CF [120], which aim at pinpointing the exact unknown tags. For fair comparison, we do not simulate the process of unknown tag collection of these two identification protocols.

6.5.2.1 Execution time

Experimental results shown in Table 6.2 demonstrate that the proposed SBF-UDP considerably outperforms all the previous related protocols. For example, when $N = 30,000$, $\lambda = 3$, and $\beta = 99\%$, the execution time of CU, BUIP-CF, SEBA, and SEBA+ is 121.0, 76.6, 49.5, and 38.5 s, respectively. And the execution time of SBF-UDP

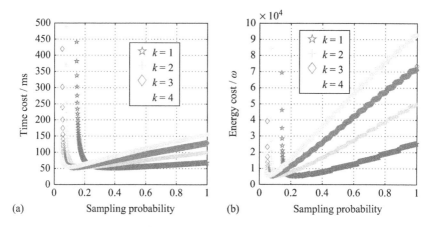

Figure 6.4   Impact of p on SBF-UTD. $N = 3,000$, $\lambda = 20$, $\beta = 95\%$. k is the
number of hashing functions. (a) Time cost vs. p and (b) energy
cost vs. p

Table 6.2   Time cost of the protocols, where N varies from 10,000 to 30,000; λ and
β are fixed to 3% and 99%

Alg. name	CU (s)	BUIP-CF (s)	SEBA (s)	SEBA+ (s)	SBF-UDP(t) (s)	SBF-UDP(e) (s)
$N = 10,000$	40.4	25.6	16.5	12.8	1.3	1.7
$N = 15,000$	60.5	37.9	24.7	19.2	2.0	2.6
$N = 20,000$	80.7	51.4	32.9	25.6	2.7	3.4
$N = 25,000$	101.3	64.0	41.2	32.0	3.3	4.3
$N = 30,000$	121.0	76.6	49.5	38.5	4.0	5.2

working in time-saving mode is 4.0 s, which represents 30 times faster than CU,
19 times faster than BUIP-CF, 12 times faster than SEBA, and 9 times faster than
SEBA+.

6.5.2.2   Energy cost

The simulation results shown in Table 6.3 reveal that the proposed SBF-UDP significantly reduces the energy consumption compared with the previous protocols. For
example, when $N = 30,000$, $\lambda = 3$, and $\beta = 99\%$, the energy cost of CU, BUIP-CF,
SEBA, and SEBA+ is $4.52 \times 10^8 \omega$, $1.13 \times 10^8 \omega$, $1.86 \times 10^8 \omega$, and $1.92 \times 10^8 \omega$,
respectively. While the energy cost of SBF-UDP working in energy-saving mode is
$1.94 \times 10^7 \omega$, reducing the energy consumption by 95.7%, 82.8%, 89.6%, and 89.9%,
respectively. The lifetime of active RFID tags could be significantly prolonged.

Table 6.3 *Energy cost of the protocols, where N varies from 10,000 to 30,000; λ and β is fixed to 3% and 99%*

Alg. name	CU (ω)	BUIP-CF (ω)	SEBA (ω)	SEBA+ (ω)	SBF-UDP(t) (ω)	SBF-UDP(e) (ω)
$N = 10{,}000$	5.10×10^7	1.25×10^7	2.06×10^7	2.14×10^7	2.23×10^6	2.17×10^6
$N = 15{,}000$	1.13×10^8	2.85×10^7	4.64×10^7	4.80×10^7	5.00×10^6	4.85×10^6
$N = 20{,}000$	2.02×10^8	5.05×10^7	8.24×10^7	8.55×10^7	8.92×10^6	8.62×10^6
$N = 25{,}000$	3.15×10^8	7.85×10^7	1.29×10^8	1.34×10^8	1.39×10^7	1.35×10^7
$N = 30{,}000$	4.52×10^8	1.13×10^8	1.86×10^8	1.92×10^8	2.00×10^7	1.94×10^7

6.5.3 *The actual detection reliability*

In this subsection, we conduct simulations to evaluate the actual detection reliability of the proposed SBF-UDP. The number N of the known tags varies from $1{,}000$ to $10{,}000$. The tolerance number λ and the detection reliability β is set to 10% and 95%, respectively. We simulated different numbers M of unknown tags hiding in the system, where M varies from 10 to 13. The simulation results shown in Figure 6.5 demonstrate that the proposed SBF-UDP protocol can meet the predefined detection reliability under both time-saving mode and energy-saving mode. Specifically, the actual detection reliability fluctuates around the predefined reliability 95% when exactly $\lambda = 10$ unknown tags are simulated in Figure 6.5(a), which is the most difficult case for the detection. This fluctuation because of the probabilistic variance is reasonable and acceptable in practice. When more than λ tags are simulated hiding in the system, e.g., Figure 6.5(b)–(d), the actual detection reliability is usually larger than the predefined reliability 95%.

6.5.4 *The impact of channel error*

For the clarity of description, this chapter assumed (so far) a perfect communication channel between the reader and tags [98,108]. We now propose a countermeasure to mitigate negative impact of channel error on our SBF-UTD protocol. The f-bit BF is divided into multiple segments, each of which is 80 bits. The computer calculates CRC-16 (cyclic redundancy check) of each segment. The reader sequentially broadcasts the binary $\langle$segment, CRC$\rangle$ to the tags. The 96-bit binary can be transmitted by the reader in a tag slot. We assume each bit in the binary is corrupted due to channel error with a probability of P_{BER} during the transmission process. If some bits in the binary $\langle$segment, CRC$\rangle$ are corrupted, the segment and the CRC confined in a binary will not match. If a tag finds the received $\langle$segment, CRC$\rangle$ is corrupted, it will not use the segment to check whether it is unknown tag. A simple but effective countermeasure to mitigate the impact of channel error is to broadcast $c > 1$ copies of $\langle$segment, CRC$\rangle$. Multiple copies can increase the probability that at least one of them is correctly transmitted. With fixed N, λ and β, the time and energy costs will be proportional to the number of copies transmitted. The simulation results

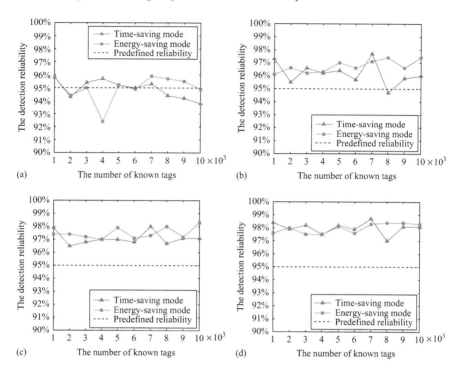

Figure 6.5   *Evaluating the actual detection reliability of the proposed SBF-UDP,*
where N varies from 1,000 to 10,000; $\lambda = 10$; $\beta = 95\%$. (a) $\lambda = 10$
unknown tag is simulated; (b) $\lambda + 1 = 11$ unknown tags are simulated;
(c) $\lambda + 2 = 12$ unknown tags are simulated; (d) $\lambda + 3 = 13$ unknown tags
are simulated

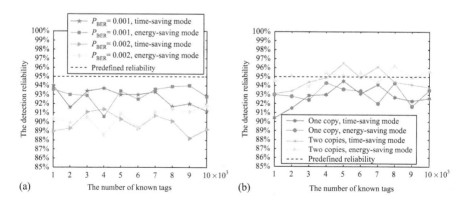

Figure 6.6   *Evaluating the impact of channel error on the accuracy of SBF-UTD,*
where N varies from 1,000 to 10,000. $\lambda = 10$, $\beta = 95\%$. 10 unknown
tags are simulated. (a) Investigating the impact of different P_{BER}.
(b) Investigating the effectiveness of multicopy countermeasure

in Figure 6.6(a) show that the channel error decreases the detection accuracy, and the results in Figure 6.6(b) demonstrate that our multicopy countermeasure is able to mitigate the negative impact of channel error.

6.6 Conclusion

This study has investigated a practically important problem of unknown tag detection that aims to detect the existence of unknown tags in a time- and energy-efficient way. We have proposed a new SBF which is a general case of the standard BF. Based on the new filtering technique, we have further proposed the SBF-UDP. Sufficient theoretically analysis is presented to minimize the execution time as well as energy consumption of the proposed protocol. We conduct extensive simulations to evaluate the performance of our SBF-UDP, and the experimental results show that our SBF-UDP protocol considerably outperforms the previous related protocols in terms of both time efficiency and energy efficiency.

Chapter 7
RFID queries—single category

7.1 Introduction

7.1.1 Background and motivation

As the cost of commercial RFID tags has become negligible compared to the prices of the products to which they are attached [24], RFID systems have been increasingly used in various applications such as supply chain management [15], indoor localization [20], inventory control, and access control [8]. For example, Walmart uses RFID tags to track expensive clothing merchandize [33], and Honeywell Aerospace uses RFID tags to track its products from birth to repair and retirement [34]. An RFID system consists of tags and readers. A tag is a microchip with an integrated antenna in a compact package that has limited computing power and communication range. There are two types of tags: passive tags and active tags. Passive tags do not have their own power source and are powered up by harvesting the radio frequency energy from readers, while the active tags have their own power sources. A reader has a dedicated power source with a significant amount of computing power. RFID systems work in a query–response fashion where a reader transmits queries to a set of tags and the tags respond with their IDs over a shared wireless medium.

This chapter addresses the fundamental problem of RFID tag searching: *given a set of known tag IDs and a population of RFID tags with unknown IDs, where the tags may be passive or active, we want to know which tag IDs are in the tag population, i.e., search in a population of unknown tags for a set of known IDs*. RFID tag searching finds applications in product recall, inventory balancing, stock verification, and many other such settings. For product recall, if a manufacturer suspects that some of its products, which have already been distributed in different warehouses, are defective, they can use a tag searching protocol to quickly locate defective products, where the known tag IDs are defective products and the tag population are the products in a warehouse. For inventory balancing, if a large retailer, such as Amazon, wants to balance the quantity of different products among its warehouses across the country to reduce shipping time and costs, they can use a tag searching protocol to determine the quantity of any given product in each warehouse and then balance the quantity among warehouses accordingly, where the known tag IDs are the ones in inventory and the tag population are the ones in a warehouse. For stock verification, if a large retailer wants to check the quantity of each requested product sent to it in a large consignment, they can use a tag searching protocol to determine whether the consignment contains all requested products, where the known tag IDs are the ones that they are expecting and the tag population are the ones in the consignment. In this chapter, we use the three terms, a tag, a tag ID, and the product that a tag is attached to, interchangeably.

7.1.2 Problem statement

Now, we formally define the tag searching problem. *Given a set A, which is a set of known tag IDs, a set B, which is a population of RFID tags with unknown IDs, a required confidence interval β, a tag searching protocol outputs $\tilde{C}$ so that $C \subseteq \tilde{C} \subseteq A$ and $|\tilde{C}| - |C| \leq \beta |C|$, where $C = A \cap B$.* Confidence interval β represents the maximum tolerable fraction of tags in A that are not in C but are declared as members of C by a tag searching protocol. A tag searching protocol should satisfy three additional requirements. First, it should comply with the EPCGlobal Class 1 Generation 2 (C1G2) RFID standard [53], which is a stable RFID standard and followed by the commercial RFID devices. Otherwise, it will be extremely difficult to be practically deployed. Second, it should preserve the privacy of the RFID tags in set B by not reading their tag IDs. Many RFID tag searching applications need to satisfy this privacy requirement. For example, if a policeman searches for some items with known tag IDs in a private house with a population of tags with unknown tag IDs, the home owner may prefer not to read the IDs of all tags in the house. Third, it should work with both a single-reader and multiple-reader environments. As the communication range between a tag and a reader is limited, a large population of tags is often covered by multiple readers with overlapping regions.

7.1.3 Limitations of prior art

Previous RFID tag searching protocols (i.e., [81,103,128]) have two key limitations. First, they cannot achieve arbitrarily high accuracy. They are all probabilistic in nature, but none of them takes the confidence interval β as an input. Second, they do not comply with the C1G2 standard as they require the tags to receive, interpret, and act either according to preframe bloom filters or other protocol specific parameters. It is critical for RFID protocols to be compliant with the C1G2 standard because the cheap commercially available off-the-shelf (COTS) tags follow the C1G2 standard. A protocol that does not comply with the C1G2 standard will require custom tags, which will cost significantly more and have limited applications. Previous RFID identification protocols (such as TH [44], STT [21], MAS [17], and ASAP [22]) can be used to read all IDs of the tags in B and then calculate $C = A \cap B$. However, this straightforward solution has two key limitations. First, it does not preserve the privacy of the tags in B as it needs to read the IDs of all tags in B. Second, this is inefficient. We want an RFID tag searching protocol that is much faster than reading all tags in B.

7.1.4 Proposed approach

In this chapter, we propose a protocol called RFID tag searching protocol (RTSP), which satisfies the following four requirement: (1) C1G2 compliance, (2) arbitrary accuracy, i.e., $C \subseteq \tilde{C} \subseteq A$ and $|\tilde{C}| - |C| \le \beta|C|$ for any required confidence interval β, (3) privacy preserving, and (4) multiple-reader capability.

To satisfy the requirement of C1G2 compliance, RTSP uses the framed-slotted Aloha protocol specified in the C1G2 standard as its MAC layer communication protocol. In Aloha, the reader first tells the tags a frame size f and a random seed number R. Each tag within the transmission range of the reader then uses f, R, and its ID to select a slot in the frame by calculating a hash function $h(f, R, \text{ID})$ whose result is uniformly distributed in $[1, f]$. Each tag has a counter initialized with the slot number that it chose to reply. After each slot, the reader first transmits an end of slot signal and then each tag decrements its counter by one. In any given slot, all the tags whose counters equal 1 respond with a random sequence called RN16. The reader uses this sequence to determine whether one or more than one tags are replying in that slot. If no tag replies in a slot, it is called an *empty slot*. If one or more tags reply in a slot, it is called a *nonempty slot*. Using 0 to denote an empty slot and 1 to denote a nonempty slot, after we execute the Aloha protocol on a population A of tags using frame size f and random seed R, we obtain a binary array of f bits, denoted as $\mathbb{S}(A, f, R)$.

To satisfy the requirement of arbitrary accuracy, RTSP executes n runs of the Aloha protocol where each run uses a different seed. For the ith run with frame size f and random seed R_i, RTSP executes the Aloha protocol on both sets A and B, and thus obtains two binary arrays $\mathbb{S}(A, f, R_i)$ and $\mathbb{S}(B, f, R_i)$. Note that RTSP executes the Aloha protocol on A virtually as it knows all tag IDs in A. After n runs, for each tag ID $t \in A$, if for all $1 \le i \le n$, we have $\mathbb{S}(A, f, R_i)[h(f, R_i, t)] = \mathbb{S}(B, f, R_i)[h(f, R_i, t)]$, (i.e., for all n runs, the two bits corresponding to tag t in both $\mathbb{S}(A, f, R_i)$ and $\mathbb{S}(B, f, R_i)$ are 1), then RTSP outputs $t \in \tilde{C}$. Clearly, RTSP satisfies $C \subseteq \tilde{C} \subseteq A$. RTSP chooses a value of n so that $|\tilde{C}| - |C| \le \beta|C|$.

To satisfy the requirement of privacy preserving, RTSP checks if a slot is empty or nonempty using the RN16 sequence and never asks tags to transmit their IDs. In C1G2, tags do not transmit their IDs unless the reader specifically asks them.

To satisfy the requirement of multireader capability, RTSP uses a central controller for all readers to use the same values for frame size f and seed R across all readers. The central controller uses a reader scheduling protocol [27] to ensure that two readers with overlapping regions do not transmit at the same time. When a reader transmits seed R_i in its ith frame, it does not generate R_i on its own, rather, it uses the ith seed R_i issued by the central controller. Thus, for a tag $t \in B$ that is covered by multiple readers, it chooses the same slot $h(f, R_i, t)$ for all readers. Once a reader completes its frame, it sends its binary array to the central controller. The controller applies the bit-wise logical OR operation on the binary arrays returned from all readers. The resulting binary array is the same as if there is one reader that covers all tags. RTSP uses this binary array to compute $\tilde{C}$.

7.1.5 Technical challenges and proposed solutions

There are two key technical challenges in RTSP. The first technical challenge is to minimize tag searching time under the constraint that RTSP satisfies the required accuracy. To address this challenge, we use the accuracy requirement $|\tilde{C}| - |C| \leq \beta|C|$ to derive a *confidence condition*, which the system parameters such as frame sizes and execution rounds must satisfy. We then use the confidence condition to derive a *duration condition*, which system parameters must satisfy to minimize tag searching time. We then solve both conditions simultaneously to calculate the optimum system parameters that minimize tag searching time while achieving the required accuracy.

The second technical challenge is to estimate the number of tags in set $|C|$, which is required to calculate the optimal values of system parameters. To address this challenge, RTSP counts the number of bits that are 1s in both $\mathbb{S}(A, f, R_i)$ and $\mathbb{S}(B, f, R_i)$. We call such bits dual-nonempty bits. The number of such dual-nonempty bits is a monotonically increasing function of $|C|$. By observing the number of dual-nonempty slots, RTSP estimates the value of $|C|$. In this chapter, we estimate $|C|$ while executing the Aloha protocol.

7.1.6 Advantages over prior art

The key novelty of this chapter is in proposing a tag searching protocol that statistically guarantees to achieve any required accuracy and complies with the C1G2 standard. The key technical depth of RTSP lies in its mathematical development to guarantee any required accuracy and to minimize tag searching time. The key advantages of RTSP over prior tag searching protocols are that RTSP can achieve arbitrarily high accuracy and RTSP complies with the C1G2 standard. RTSP is easy to deploy because it is implemented on readers as a software module and does not require any implementation on tags. Furthermore, it does not require any modifications either to tags or to the communication protocol between tags and readers and works with the commercially available off-the-shelf RFID tags. RTSP can be implemented as a software module on readers. We have extensively evaluated the performance of RTSP. Our results show that for a scenario with $|A| = 5,000$, $|B| = 5,000$, and $|C| = 500$, and a required confidence interval of 0.1%, RTSP takes 15 s to search the tags, whereas the fastest prior tag identification protocol (TH [44]) takes 22 s.

7.2   Related work

To the best of our knowledge, there are four tag searching protocols [81,103,128,129]. Zheng and Li proposed the first RFID tag searching protocol namely CATS [81]. CATS works in two phases. In the first phase, a server first constructs a bloom filter by applying multiple hash functions in conjunction with a random seed on each tag ID in set A. Second, an RFID reader broadcasts the bloom filter generated by the server along with the random seed to all tags in the population B. Using the received bloom filter of set A, each tag in B checks if it is a candidate in C. Specifically, if all bits for a tag are 1s, the tag is a candidate in C; otherwise, it must be in $B - A$. Let B' denote all these candidates. Thus, due to false positives, $C \subseteq B' \subseteq B$. Then, the tags in B' distributively construct another bloom filter using the framed-slotted Aloha protocol. The reader uses this bloom filter to exclude the IDs in $A - B$. Thus, the reader obtains the searching result $A - (A - B) = A \cap B$. Unfortunately, C1G2 compliant tags cannot interpret or generate bloom filters, which makes CATS noncompliant with the C1G2 standard.

Chen *et al.* proposed another tag searching protocol called ITSP, which is an improved version of CATS [103]. In ITSP, the reader first generates a $k = 1$ bloom filter on set A. Then, the reader broadcasts the bloom filter along with the parameters used for constructing the bloom filter to all tags in B. After a tag receives the bloom filter, it checks whether it is in the bloom filter. If a tag is in the bloom filter, the tag will remain active; otherwise, it will become inactive. For the active tags, they collaboratively construct another $k = 1$ bloom filter by executing the framed-slotted Aloha protocol. ITSP repeats the above filtering process for multiple rounds until the false positive probability is below a certain threshold. Unfortunately, C1G2 compliant tags cannot interpret or generate bloom filters, which makes ITSP noncompliant with the C1G2 standard.

Zhang *et al.* proposed another tag searching protocol called TSM [128]. TSM extends CATS for use with multiple readers. It first executes CATS using each reader and then aggregates results from all readers to identify the tags in A that are present in B. Unfortunately, due to similar reasons as for CATS, TSM is also noncompliant with the C1G2 standard. In contrast, our proposed protocol, RTSP, is C1G2 compliant.

Liu *et al.* proposed BKC to count the number of tags in A that are present in B [129]. BKC first precomputes a frame using IDs in set A and then executes a frame on population B to determine how many times the slots that were 1 in the precomputed frame turned out to be 1 in the executed frame. It then uses the number of such slots to obtain the estimate of the number of tags in A that are present in B. BKC falls short because it can only estimate the number of tags in A that are present in B, but it cannot determine exactly which tags of A are present in B. In contrast, our proposed protocol RTSP can identify such tags.

7.3 System model

7.3.1 Architecture

For searching RFID tags, RTSP uses a central controller connected with a set of readers that cover the area where the tags in set B are located. The use of a central controller ensures that all readers use consistent values of frame sizes and seeds when executing frames, which helps in efficiently aggregating and processing information returned by the readers. The readers use the standardized framed-slotted Aloha protocol to communicate with tags and never ask the tags to transmit their IDs. The use of multiple readers with overlapping coverage regions introduces following two problems: (1) scheduling the readers such that no two readers with overlapping regions transmit at the same time, and (2) alleviating the effect of some tags responding to multiple readers due to overlap in the coverage region of those readers. For the first problem, the controller uses one of the several existing reader scheduling protocols [27] to avoid reader–reader collisions. For the second problem, we propose solution in Section 7.4.1. RTSP does not require any modifications to tags or readers. It only requires the readers to receive system parameters from the controller and communicate the responses in the frames back to the controller.

7.3.2 C1G2 compliance

RTSP does not require any modifications to tags or readers. It only requires the readers to receive the frame size, persistence probability, and seed number from the controller and communicate the responses in the frames back to the controller. Persistence probability p is the probability with which a tag decides whether it will participate in a frame or not before selecting a slot in that frame. Later, in the chapter, we will show how we use p to handle frame sizes that exceed the C1G2 specified upper limit of 2^{15}. Such large frame sizes are required when the size of tag population is large and required confidence interval β is small. With the use of p, the reader reduces the number of tags that participate in each frame, which in turn reduces the optimal frame size at the expense of increased number of frames. As the C1G2 standard does not specify the use of p, COTS tags do not support it. To avoid making any modifications to tags, in RTSP, the reader implements p by announcing a frame size of f/p but terminating the frame after the first f slots and sending a command to tags to reset their counters, which can be done as per the C1G2 standard.

7.3.3 Communication channel

We assume that the communication channel between readers and tags is reliable i.e., tags correctly receive queries from the readers and the readers correctly detect transmission of RN16 sequence in a slot if one or more tags in the population transmit in that slot. If the channel is unreliable, the solution proposed in [44] can be easily adapted for use with RTSP.

7.3.4 Independence assumption

To make the formal development tractable, we assume that instead of picking a single slot to transmit at the start of ith frame of size f, a tag independently decides to transmit in each slot of the frame with probability $1/f$ regardless of its decision about previous or forthcoming slots. Vogt first used this assumption for the analysis of Aloha protocol for RFID and justified its use by recognizing that this problem belongs to a class of problems called *occupancy problem*, which deals with the allocation of balls to urns [49]. Ever since, the use of this assumption has become a norm in the formal analysis of all Aloha-based RFID protocols [30,45,49].

The implication of this assumption is that a tag can end up choosing more than one slots in the same frame or even not choosing any at all, which is not in accordance with the C1G2 standard that requires a tag to pick exactly one slot in a frame. However, this assumption does not create any problems because the expected number of slots that a tag chooses in a frame is still one. The analysis with this assumption is, therefore, asymptotically the same as that without this assumption [51]. Bordenave *et al.* further explained in detail why this independence assumption in analyzing

Aloha-based protocols provides results just as accurate as if all the analysis was done without this assumption [51]. This independence assumption is made only to make the formal development tractable. In our simulations, tag chooses exactly one slot at the start of frame.

7.4 RFID tag search protocol

7.4.1 Protocol description

To search which tags in set A are present in the population B, in RTSP, the central controller executes n Aloha frames using the RFID readers. There are five steps involved in executing each frame. First, before executing any frame i, the controller calculates the optimal values of frame size f_i, persistence probability p_i, and generates a random seed number R_i. We will derive the expressions to calculate the values of f_i and p_i in the next section. Second, as the controller knows the IDs in set A, it virtually executes the Aloha protocol on set A and obtains the binary array $\mathbb{S}(A, f_i, R_i)$. Thus, the controller knows which bits in the binary array $\mathbb{S}(B, f_i, R_i)$ resulting from executing ith frame on population B should be 1 if all the tags in A were present, and a single reader covered the entire population. Third, it provides each reader with the parameters f_i, p_i, and R_i and asks each of them to execute the ith frame using these parameters. The motivation behind using the same values of f_i, p_i, and R_i across all readers for the ith frame is to enable RTSP to work with multiple readers with overlapping regions. As all readers use the same values of f_i, p_i, and R_i in the ith frame, the slot number that a particular tag chooses in the ith frame of each reader covering this tag is the same i.e., $h(\frac{f_i}{p_i}, R_i, \text{ID})$ evaluated by the tag results in same value for each reader. Fourth, each reader executes the frame on its turn as per the reader scheduling protocol and sends the responses in the frame back to the controller. Fifth, after the controller has received the ith frame of each reader, it applies logical OR operator on all the received ith frames and obtains the resultant bit array $\mathbb{S}(B, f_i, R_i)$. This resultant bit array $\mathbb{S}(B, f_i, R_i)$ is the same as if generated by a single reader covering all the tags. After obtaining the n bit arrays, $\mathbb{S}(B, f_i, R_i)$ for $1 \leq i \leq n$, for each tag $t \in A$, if $h(\frac{f_i}{p_i}, R_i, t) \leq f_i$ the controller checks whether $\mathbb{S}(A, f_i, R_i)[h(\frac{f_i}{p_i}, R_i, t)] = \mathbb{S}(B, f_i, R_i)[h(\frac{f_i}{p_i}, R_i, t)]$ for all n frames, i.e., for all n frames, whether the two bits corresponding to tag t in both $\mathbb{S}(A, f_i, R_i)$ and $\mathbb{S}(B, f_i, R_i)$ are 1s. If true, RTSP declares that the tag t is present in B. Note that RTSP can have false positives, i.e., it can declare a tag in set A to be present in population B, when the tag is actually not present. RTSP does not have false negatives.

7.4.2 Estimating number of tags in set C

Recall from the previous section that before executing any frame i, the controller calculates the optimal values of frame size f_i and persistence probability p_i. To calculate these optimal values for ith frame, the controller needs estimate of $|C|$ at start of the ith frame, which it obtains using the responses from the tag population in the previous $i - 1$ frames. We represent the estimate of $|C|$ at the start of ith frame by $|\tilde{C}_i|$.

As the controller executes more and more frames, i.e., as i increases, the estimate $|\tilde{C}_i|$ asymptotically becomes equal to $|C|$. Next, we present a method to estimate the value of $|C|$ at start of any frame i.

The intuition behind our estimation method is that as the number of tags in set C increases, the number of corresponding bits that are 1s in both $\mathbb{S}(A, f_i, R_i)$ and $\mathbb{S}(B, f_i, R_i)$ also increases. We call such bits as dual-nonempty bits. The number of dual-nonempty bits for any given frame is a function of $|C|$ and can, therefore, be used to estimate the value of $|C|$. Next, we derive an expression that relates the number of dual-nonempty bits with the value of $|C|$, i.e., we derive an expression for $E[\mathcal{N}_i^{11}]$ as a function of $|C|$, where $\mathcal{N}_i^{11}$ is random variable for number of dual-nonempty bits in the pair of arrays $\mathbb{S}(A, f_i, R_i)$ and $\mathbb{S}(B, f_i, R_i)$. To derive the expression for $E[\mathcal{N}_i^{11}]$, we need the probability that any given pair of bits in the arrays $\mathbb{S}(A, f_i, R_i)$ and $\mathbb{S}(B, f_i, R_i)$ is dual-nonempty. We calculate this probability in the following lemma.

Lemma 7.1. *Let A be the set of IDs of tags that we want to search for in a population. Let B be the set of IDs of tags in the population in which we search for tags in set A. Let C be the set of IDs of those tags that are present in both sets A and B. Let X_{ij} be an indicator random variable for the event that the jth bit in ith pair of arrays is a dual-nonempty bit. For frame size f_i and persistence probability p_i, the probability distribution of X_{ij} is given by the following equation:*

$$P\{X_{ij} = 1\} = 1 - \left(1 - \frac{p_i}{f_i}\right)^{|A|} - \left(1 - \frac{p_i}{f_i}\right)^{|B|} + \left(1 - \frac{p_i}{f_i}\right)^{|A|+|B|-|C|} \quad (7.1)$$

Proof. Probability that any given bit j in a pair of arrays is a dual-nonempty bit can be obtained by first calculating the probability that this bit is not a dual-nonempty bit, and then subtracting it from 1. The jth bit is not dual-nonempty when one of the following three cases happens.

1. None of the tags in set A select the jth slot in frame i.e., the jth bit in $\mathbb{S}(A, f_i, R_i)$ is 0, and none of the tags in population B selects the jth slot in corresponding executed frame i.e., the jth bit in $\mathbb{S}(B, f_i, R_i)$ is 0. We represent this event by an indicator random variable Y_{00}. The probability distribution of Y_{00} is given by the following equation:

$$P\{Y_{00} = 1\} = \left(1 - \frac{p}{f}\right)^{|A|+|B|-|C|} \quad (7.2)$$

2. One or more tags in set $A - C$ select the jth slot in frame i.e., the jth bit in $\mathbb{S}(A, f_i, R_i)$ is 1, and none of the tags in population B selects the jth slot in corresponding executed frame i.e., the jth bit in $\mathbb{S}(B, f_i, R_i)$ is 0. We represent this event by an indicator random variable Y_{10}. The probability distribution of Y_{10} is given by the following equation:

$$P\{Y_{10} = 1\} = \left(1 - \left(1 - \frac{p}{f}\right)^{|A-C|}\right)\left(1 - \frac{p}{f}\right)^{|B|} \quad (7.3)$$

3. None of the tags in set A selects the jth slot in frame i.e., the jth bit in $\mathbb{S}(A, f_i, R_i)$ is 0, and one or more tags in population $B - C$ select the jth slot in corresponding executed frame, i.e., the jth bit in $\mathbb{S}(B, f_i, R_i)$ is 1. We represent this event by an indicator random variable Y_{01}. The probability distribution of Y_{01} is given by the following equation:

$$P\{Y_{01} = 1\} = \left(1 - \left(1 - \frac{p}{f}\right)^{|B-C|}\right)\left(1 - \frac{p}{f}\right)^{|A|} \tag{7.4}$$

The distribution of X_{ij} is given by the following equation:

$$P\{X_{ij} = 1\} = 1 - P\{Y_{00} = 1\} - P\{Y_{10} = 1\} - P\{Y_{01} = 1\} \tag{7.5}$$

Substituting the expressions for the probability distributions of Y_{00}, Y_{10}, and Y_{01} from (7.2) to (7.4), respectively, into (7.5) and simplifying, we get (7.1). $\square$

Following theorem derives the expression for $E[\mathcal{N}_i^{11}]$ as a function of $|C|$.

Theorem 7.1. *Let A be the set of IDs of tags that we want to search for in a population. Let B be the set of IDs of tags in the population in which we search for tags in set A. Let C be the set of IDs of those tags that are present in both sets A and B. Let $\mathcal{N}_i^{11}$ be the random variable for the number of dual-nonempty bits in a pair of arrays of size f_i each. When persistence probability is p_i, the expected value of $\mathcal{N}_i^{11}$ is given by the following equation:*

$$E[\mathcal{N}_i^{11}] = f_i \times \left(1 - \left(1 - \frac{p_i}{f_i}\right)^{|A|} - \left(1 - \frac{p_i}{f_i}\right)^{|B|}\right.$$
$$\left. + \left(1 - \frac{p_i}{f_i}\right)^{|A|+|B|-|C|}\right) \tag{7.6}$$

Proof. It is straight forward to see that $\mathcal{N}_i^{11} = \sum_{j=1}^{f_i} X_{ij}$. As $\{X_{i1}, X_{i2}, \ldots, X_{if_i}\}$ forms a set of identically distributed random variables, $E[\mathcal{N}_i^{11}]$ is given by:

$$E[\mathcal{N}_i^{11}] = E\left[\sum_{j=1}^{f_i} X_{ij}\right] = f_i \times E[X_{ij}] \tag{7.7}$$

As expected value of an indicator random variable equals its probability of being 1, $E[X_{ij}] = P\{X_{ij} = 1\}$. Substituting value of $E[X_{ij}]$ in equation above with value of $P\{X_{ij} = 1\}$ from (7.5), we get the equation for $E[\mathcal{N}_i^{11}]$. $\square$

Figure 7.1 plots $E[\mathcal{N}_i^{11}]$ as a function of $|C|$ using (7.6). This figure is obtained using $|A| = 200$, $|B| = 300$, $f_i = 300$, and $p_i = 1$. We observe from this figure that $E[\mathcal{N}_i^{11}]$ is a monotonically increasing function of $|C|$.

To estimate the value of $|C|$, let $\tilde{\mathcal{N}}_i^{11}$ represent the observed value of number of dual-nonempty bits for ith pair of bit arrays. Replacing $E[\mathcal{N}_i^{11}]$ in (7.6) with $\tilde{\mathcal{N}}_i^{11}$ and

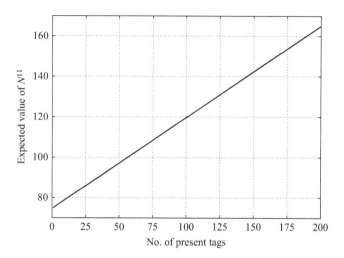

Figure 7.1 $E[\mathcal{N}_i^{11}]$ vs. $|C|$

solving for $|C|$ gives an estimate of $|C|$. Using the well-known identity $(1 + x)^y \approx e^{xy}$ for small x and large y, (7.6) can be written as follows:

$$E[\mathcal{N}_i^{11}] \approx f_i \times \left(1 - e^{-\frac{p_i}{f_i}|A|} - e^{-\frac{p_i}{f_i}|B|} + e^{-\frac{p_i}{f_i}(|A|+|B|-|C|)}\right) \tag{7.8}$$

Replacing $E[\mathcal{N}_i^{11}]$ in the equation above with $\tilde{\mathcal{N}}_i^{11}$ and solving for $|C|$, we get the following equation to obtain the estimate $|\tilde{C}|$ of $|C|$:

$$|\tilde{C}| \approx |A| + |B| + \frac{f_i}{p_i}\ln\left\{\frac{\tilde{\mathcal{N}}_i^{11}}{f_i} - 1 + e^{-\frac{p_i}{f_i}|A|} + e^{-\frac{p_i}{f_i}|B|}\right\} \tag{7.9}$$

This estimate is obtained by utilizing the information from the ith frame only. While this estimate may not be accurate, if we use the information from more frames, the estimate will become more accurate. Specifically, we leverage the well-known statistical result that the variance in the observed value of a random variable reduces by x times if we take the average of x observations of that random variable. Therefore, to obtain the estimate $|\tilde{C}_i|$ of $|C|$ at the start of the ith frame, we obtain an estimate from each of the previous $i - 1$ frames and take their average. Solving (7.6) for $|C|$ and averaging over past $i - 1$ frames, the formal expression for $|\tilde{C}_i|$ becomes:

$$|\tilde{C}_i| \approx |A| + |B| + \frac{\sum_{l=1}^{i-1}\frac{f_l}{p_l}\ln\left\{\frac{\tilde{\mathcal{N}}_l^{11}}{f_l} - 1 + e^{-\frac{p_l}{f_l}|A|} + e^{-\frac{p_l}{f_l}|B|}\right\}}{i - 1} \tag{7.10}$$

Note that $|B|$ can be obtained using existing RFID estimation schemes such as ART [45]. Further note that the controller obtains this estimate without executing any additional frames. It gets this estimate from the frames it was already executing to search for tags.

7.5 Parameter optimization

In this section, we will derive equations that the controller uses at the start of ith frame to calculate the optimal values of frame size f_i and persistence probability p_i to minimize the execution time of RTSP while ensuring that its actual confidence interval is less than the required confidence interval. At the start of ith frame, the controller uses the estimate $|\tilde{C}_i|$ along with the values of $|A|$, $|B|$, and β to calculate the optimal values of f_i and p_i. Before asking the readers to execute the ith frame, the controller also calculates the minimum number of frames that it should execute, represented by n_i. Recall from Section 7.4.2 that as the number of executed frames increase, the estimate of $|C|$ becomes more accurate. Consequently, n_i, f_i, and p_i asymptotically become equal to constants n, f, and p, respectively. When the estimate of $|C|$ changes by less than 2 in 10 consecutive frames, the controller considers the estimate to be close enough to $|C|$. At this point, the controller calculates the values of n_i, f_i, and p_i one last time and puts $f = f_i$, $p = p_i$, and $n = n_i$, and uses these fixed values of f and p to execute subsequent frames until the total number of frames executed since the first frame becomes equal to n. For the first frame, i.e., when $i = 1$, the controller uses $n_1 = \infty$, $f_1 = \max\{|A|, |B|\}$, and $p_1 = 1$. The choices of the values of n_1, f_1, and p_1 are arbitrary and do not really matter because as the controller executes more frames, number of frames, frame size, and persistence probability converge to constants n, f, and p, respectively.

In subsequent calculation of n_i, f_i, and p_i, we will drop the subscript i to make the presentation simple. Next, we first derive the expression for false positive probability i.e., probability with which RTSP declares a tag in set A to be present in population B, when it actually is not. Second, using the expression for false positive probability, we derive a *confidence condition* that the values of n, f, and p must satisfy to ensure that the observed confidence interval is smaller than the required confidence interval β, i.e., the requirement $|\tilde{C}| - |C| \leq \beta|C|$ is satisfied. Third, we derive a *duration condition*, which the values of f and p must satisfy to ensure that the execution time of RTSP is minimized. The controller solves these two conditions simultaneously to obtain the optimal values of n, f, and p. Last, we describe our strategy to bring the value of f within limit when the optimal frame size exceeds the C1G2-specified upper limit of 2^{15}.

7.5.1 False positive probability

A false positive occurs when all the bits that a particular tag in A that is not present in B selects in the n bit arrays $\mathbb{S}(A, f_i, R_i)$ for $1 \leq i \leq n$, turn out to be nonempty in corresponding bit arrays $\mathbb{S}(B, f_i, R_i)$ because some other tags in the population made those bits 1. Lemma 7.2 gives the expression to calculate the false positive probability.

Lemma 7.2. *Let B be the set of IDs of tags in the population in which we search for tags. With persistence probability p, frame size f, and number of frames n, the false positive probability, P_{fp}, is given by* $P_{\mathrm{fp}} = \left[1 - \left(1 - \frac{p}{f}\right)^{|B|} \right]^{n}$.

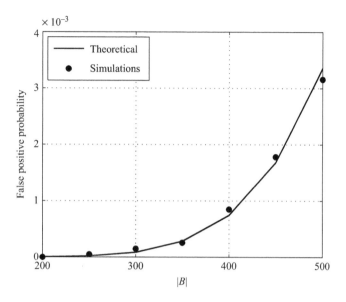

Figure 7.2   Theoretical vs. experimental P_{fp}

Proof. Consider a tag t such that $t \in A \wedge t \notin B$. The probability that the bit tag t selects in $\mathbb{S}(A, f_i, R_i)$ is selected by at least one tag in population B in $\mathbb{S}(B, f_i, R_i)$ is $1 - (1 - \frac{p}{f})^{|B|}$. The probability that all n bits tag t selects in the n bit arrays $\mathbb{S}(A, f_i, R_i)$ for $1 \le i \le n$, are also selected by some other tags in population B in corresponding bit arrays $\mathbb{S}(B, f_i, R_i)$ is $[1 - (1 - \frac{p}{f})^{|B|}]^n$, which is the expression for P_{fp}, given in the lemma statement. $\square$

Figure 7.2 shows the theoretically calculated false positive probability from equation of P_{fp} in Lemma (7.2) represented by the solid line and experimentally observed values of false positive probability represented by the dots. To obtain this figure, we use $f = 600$, $p = 1$, and $n = 10$. Each dot represents the false positive probability calculated from 200 runs of simulation. We observe that the theoretically calculated values match perfectly with experimentally observed values, showing that our independence assumption that we stated in Section 7.3.4 does not cause the theoretical analysis to deviate from practically observed values.

7.5.2   Confidence condition

Theorem 7.2 states the confidence condition, which the values of n, f, and p must satisfy to achieve the required confidence interval β.

Theorem 7.2. *Let A be the set of IDs of tags that we want to search for in a population. Let B be the set of IDs of tags in the population in which we search for tags in set A. Let C be the set of IDs of those tags that are present in both sets A and B. To ensure that RTSP satisfies the requirement $|\tilde{C}| - |C| \le \beta|C|$, the controller must use the* *values for number of frames n, frame size f, and persistence probability p that satisfy the confidence condition given in the following equation:*

$$n = \frac{\ln\left(\frac{\beta \times |\tilde{C}|}{|A| - |\tilde{C}|}\right)}{\ln\left(1 - (1 - \frac{p}{f})^{|B|}\right)} \tag{7.11}$$

Proof. Let $E[|\tilde{C}|]$ represent the number of tags that RTSP declares as belonging to set C after executing n frames of size f with persistence probability p. Replacing $|\tilde{C}|$ in $|\tilde{C}| - |C| \le \beta|C|$ by $E[|\tilde{C}|]$, the confidence requirement is given by $E[|\tilde{C}|] - |C| \le \beta|C|$. Next, we derive the expression for $E[|\tilde{C}|]$. Recall from Section 7.4.1 that RTSP can have false positives, but it cannot have false negatives i.e., it will always identify the tags of A present in B, and in addition, it may also declare some tags in A that are not in B to be present in B. Thus, $E[|\tilde{C}|] = |C| + (|A - C|) \times P_{fp}$. As $C \subseteq A$, thus, $E[|\tilde{C}|] = |C| + (|A| - |C|) \times P_{fp}$. Substituting this value of $E[|\tilde{C}|]$ into the confidence requirement, we get the following equation: $|C| + (|A| - |C|) \times P_{fp} - |C| \le \beta|C|$. Substituting the value of P_{fp} from Lemma 7.2 into this equation and rearranging, we get $n \ge \frac{\ln\left(\frac{\beta \times |C|}{|A| - |C|}\right)}{\ln\left(1 - (1 - \frac{p}{f})^{|B|}\right)}$. As we do not know the exact value of $|C|$, rather we know the estimate $|\tilde{C}|$ of $|C|$, replacing $|C|$ in this equation with $|\tilde{C}|$ and using the smallest value for n allowed by the equation above to ensure that confidence requirement is always met, we get (7.11) in theorem statement. □

7.5.3 Duration condition

Theorem 7.3 states the duration condition that the values of f and p must satisfy to minimize the execution time of RTSP.

Theorem 7.3. *Let A be the set of IDs of tags that we want to search for in a population. Let B be the set of IDs of tags in the population in which we search for tags in set A. Let C be the set of IDs of those tags that are present in both sets A and B. To ensure that the execution time of RTSP is minimum, the controller must use the values for frame size f and persistence probability p that satisfy the duration condition given in the following equation:*

$$p \times |B| = f \times \left(1 - e^{\frac{p}{f}|B|}\right) \times \ln\left\{1 - e^{-\frac{p}{f}|B|}\right\} \tag{7.12}$$

Proof. Execution time is directly proportional to the total number of slots because the duration of each slot is the same, typically 300 µs for Philips I-Code RFID reader [56]. Let S represent the total number of slots. Thus, $S = f \times n$. To ensure that RTSP achieves the required confidence interval, we use the value of n from (7.11). Thus:

$$S = \frac{f \ln\left(\frac{\beta \times |\tilde{C}|}{|A| - |\tilde{C}|}\right)}{\ln\left(1 - (1 - \frac{p}{f})^{|B|}\right)} \tag{7.13}$$

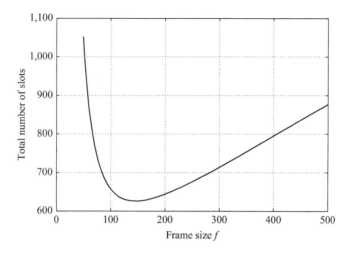

Figure 7.3   Total number of slots S vs. frame size f

Figure 7.3 plots S as a function of f using the equation above. This figure is made using $|A| = 100$, $|B| = 100$, $|\tilde{C}| = 52$, $p = 1$, and $\beta = 0.05$. We observe from this figure that S is a convex function of f. Therefore, optimum value of f exists, represented by f_{op}, that minimizes the total number of slots S. To find optimal value of f, we differentiate the equation above w.r.t. f and equate the resulting expression to 0 and get the following:

$$\left[\ln\left(\frac{\beta \times |\tilde{C}|}{|A| - |\tilde{C}|} \right) \right] \left[p|B| - f \left(1 - e^{\frac{p}{f}|B|} \right) \ln\left\{ 1 - e^{-\frac{p}{f}|B|} \right\} \right] = 0 \qquad (7.14)$$

Note that $\ln\left(\frac{\beta \times |\tilde{C}|}{|A| - |\tilde{C}|} \right) \neq 0$, which means that the following must hold true: $p|B| - f\left(1 - e^{\frac{p}{f}|B|} \right) \ln\left\{ 1 - e^{-\frac{p}{f}|B|} \right\} = 0$. Rearranging the equation above, we get the duration condition in the theorem statement.   □

The controller solves (7.11) and (7.12) simultaneously using $p = 1$ and gets the optimal values of n and f represented by n_{op} and f_{op}, respectively. It calculates f_{op} numerically from (7.12) using Brent's method. Then, it puts $f = f_{op}$ and $p = 1$ in (7.11) to calculate n_{op}. Next, we study the effect of $|A|$, $|B|$, $|C|$, and β on execution time of RTSP.

Execution time vs. $|A|$: Intuitively, as the number of tags in A increases, the execution time of RTSP should increase because the greater number of tags in A implies the higher   chances of false positives. Thus, to ensure that the number of false positives stays small enough so that the required confidence interval is achieved, RTSP executes more frames, i.e., the value of n_{op} increases, which increases the overall execution time. Figure 7.4(a) confirms our intuition. This figure plots the expected execution time of RTSP for multiple values of $|A|$ while fixing $|B|$ at 5,000 and $|C|$ at 500. We calculated the execution time as $n_{op} \times f_{op} \times T_s$, where T_s is the time of

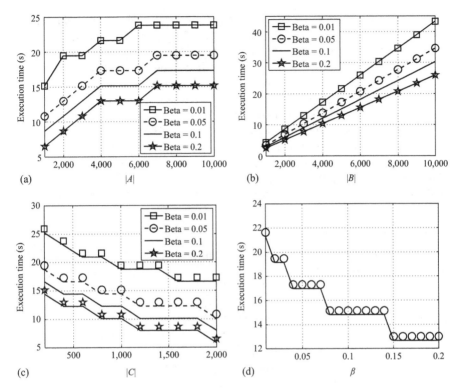

Figure 7.4 *Effect of |A|, |B|, |C|, and β on execution time of RTSP: (a) execution time vs. |A|, (b) execution time vs. |B|, (c) execution time vs. |C|, and (d) execution time vs. β*

each slot and is equal to 300 μs as per the specifications of Philips I-Code RFID reader [56]. We observe from Figure 7.4(a) that as the number of tags in A increases, the execution time of RTSP increases. The stairway behavior that RTSP shows in this and subsequent figures is due to the ceiling operation on the noninteger values of n_{op} and f_{op}.

Execution time vs. |B|: Intuitively, as the number of tags in B increases, the execution time of RTSP should increase because greater number of tags in B also imply higher chances of false positives. Thus, to ensure that the number of false positives stays small enough so that the required confidence interval is achieved, RTSP increases the frame size, i.e., the value of f_{op} increases according to (7.12), which increases the overall execution time. Figure 7.4(b) confirms our intuition. This figure plots the expected execution time of RTSP for multiple values of $|B|$ while fixing $|A|$ at 5,000 and $|C|$ at 500. We observe from Figure 7.4(b) that as the number of tags in B increases, the execution time of RTSP increases.

Execution time vs. |C|: Intuitively, as the number of tags in C increases, the execution time of RTSP should decrease because greater number of tags in C means

RTSP has greater margin of error i.e., $\beta|C|$. Thus, RTSP reduces the value of n_{op}, which decreases the overall execution time. Figure 7.4(c) confirms our intuition. This figure plots the expected execution time of RTSP for multiple values of $|C|$ while fixing $|A|$ at 5,000 and $|B|$ at 5,000. We observe from Figure 7.4(c) that as the number of tags in C increases, the execution time of RTSP decreases.

Execution time vs. β: Intuitively, as the required confidence interval β increases, the execution time of RTSP should decrease because larger required confidence interval means RTSP has greater margin of error. Thus, RTSP reduces the values of n_{op}, which decreases the overall execution time. Figure 7.4(d) confirms our intuition. This figure plots the expected execution time of RTSP for different values of β while fixing $|A|$ at 5,000, $|B|$ at 5,000, and $|C|$ at 500. We observe from Figure 7.4(d) that as the required confidence interval increases, the execution time of RTSP decreases.

7.5.4 Handling large frame sizes

For large populations and/or small required confidence interval, it is possible for the value of f_{op} to exceed the C1G2-specified upper limit of 2^{15}. Next, we describe how we use p to bring the frame size within limits. Bringing the frame size within limits comes at a cost of increased number of slots; greater than the minimum value of S that would have been achieved if the controller could use $f_{op} > 2^{15}$.

When we decrease the value of p, the number of tags that participate in a frame decreases. Therefore, the required value of f also decreases. Participation by fewer tags means that participation by the tags belonging to both the sets A and B decreases. This increases the chances that a given tag in A that is present in B will not select any slot in a given precomputed frame, which means that chances of identifying its presence decrease. Therefore, the overall uncertainty in identifying tags in A increases. To reduce this uncertainty, the n increases when p decreases to achieve the required confidence interval.

We use these two observations to reduce the value of f whenever $f_{op} > 2^{15}$. When $f_{op} > 2^{15}$, the controller uses $f = f_{max} = 2^{15}$ in (7.11), which leaves two unknowns, p and n, in the resulting equation. The controller solves the resulting equation simultaneously with (7.12) to get new values of p and n. The new value of p is less than 1, and the new value of n is greater than n_{op} (we represent n with n_{op} only when we use $f = f_{op}$ to calculate it). The controller uses these new values of n and p along with $f = f_{max}$ to compute the bit array $\mathbb{S}(A, f_i, R_i)$. Although the total number of slots $S = f_{max} \times n > f_{op} \times n_{op}$, this is still the smallest under the constraints that the required confidence interval is achieved, and the frame size does not exceed f_{max}.

7.6 Performance evaluation

We implemented and simulated RTSP in MATLAB®. We also implemented and simulated the fastest existing tag identification protocol, TH [44], to compare the execution time of RTSP with it. We choose tag ID length of 64 bits as specified in the C1G2 standard. Note that the distributions of the IDs of tags in A and B do not matter

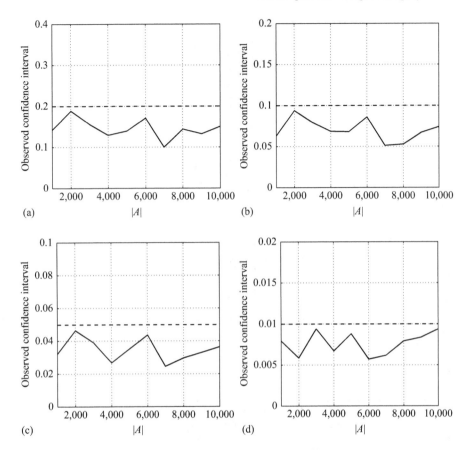

Figure 7.5 *Observed confidence interval vs.* |A| *when* |B| = 5,000, *and* |C| = 500: *(a) required β = 0.2, (b) required β = 0.1, (c) required β = 0.05, and (d) required β = 0.01*

because RTSP is independent of ID distributions. Next, we first evaluate the accuracy of RTSP and then compare its execution time with the execution time of TH. All results reported in this section are obtained from averaging over 200 independent runs of RTSP.

7.6.1 Accuracy

To evaluate the accuracy of RTSP, we study its confidence interval for different values of |A|, |B|, and |C|.

7.6.1.1 Observed confidence interval vs. |A|

Our experimental results show that RTSP always achieves the required confidence interval regardless of the size of set A. Figure 7.5(a), 7.5(b), 7.5(c), and 7.5(d) plot the actual confidence interval RTSP achieved for different sizes of set A when the

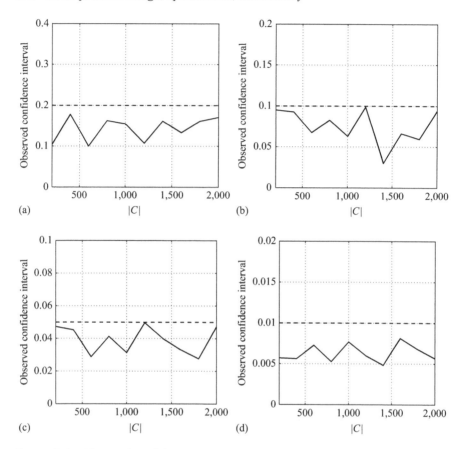

Figure 7.6   *Observed confidence interval vs. |C| when |A| = 5,000, and*
|B| = 5,000: (a) required β = 0.2, (b) required β = 0.1, (c) required
β = 0.05, and (d) required β = 0.01

required values of confidence interval are $\beta = 0.2$, $\beta = 0.1$, $\beta = 0.05$, and $\beta = 0.01$, respectively. To plot these figures, we fixed number of tags in set B at 5,000 and number of tags in A that are in B, i.e., number of tags in set C at 500. The dashed horizontal line in each of these figures shows the required value of confidence interval, and the solid line shows the observed values of confidence interval achieved by RTSP. We observe from these figures that the observed values of confidence interval are always smaller than the required values of confidence interval.

7.6.1.2   Observed confidence interval vs. |B|

Our experimental results show that RTSP always achieves the required confidence interval regardless of the number of tags in population B. We have not included figures due to lack of space.

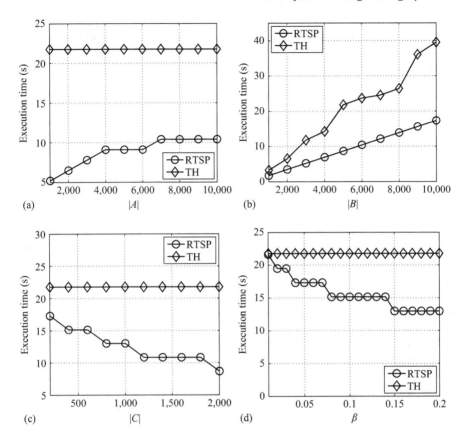

Figure 7.7 Comparison of execution times of RTSP and TH: (a) execution time vs. |A|, (b) execution time vs. |B|, (c) execution time vs. |C|, and (d) execution time vs. β

7.6.1.3 Observed confidence interval vs. $|C|$

Our experimental results show that RTSP always achieves the required confidence interval regardless of the number of tags in set C. Figure 7.6(a), 7.6(b), 7.6(c), and 7.6(d) plot the actual confidence interval RTSP achieved for different sizes of set C when the required values of confidence interval are $\beta = 0.2$, $\beta = 0.1$, $\beta = 0.05$, and $\beta = 0.01$, respectively. To plot these figures, we fixed number of tags in sets A and B at 5,000 each. Again, we observe from these figures that the solid lines are always below their corresponding dashed lines, which means that RTSP always achieves the required confidence interval.

7.6.2 Execution time

Execution time of RTSP is smaller than TH. Figure 7.7(a) plots the execution times of TH and RTSP vs. $|A|$ for $\beta = 0.1$, $|B| = 5,000$, and $C = 500$. We observe from this figure that RTSP is up to 22.73% faster compared to TH. Similarly, Figure 7.7(b) plots the execution times vs. $|B|$ for $\beta = 0.1$, $|A| = 5,000$, and $|C| = 500$, and Figure 7.7(c) plots the execution times vs. $|C|$ for $\beta = 0.1$, $|A| = 5,000$, and $|B| = 5,000$. Again, we observe from these figures that RTSP is always faster compared to TH. Finally, Figure 7.7(d) plots the execution times vs. β for $|A| = 5,000$, $|B| = 5,000$, and $|C| = 500$. We observe that RTSP is faster compared to TH as long as required confidence interval is > 0.01. When the required confidence interval < 0.01, TH is faster. Thus, if privacy is not a concern, a user should use TH whenever $\beta < 0.01$. If, however, privacy is a concern, the user should always use RTSP.

7.7 Conclusion

The key technical contribution of this chapter is in proposing a protocol to search tags in a population of RFID tags. This chapter represents the first effort on addressing this important and practical problem for C1G2 compliant RFID systems. The key technical depth of this chapter is in the mathematical development of the theory that RTSP is based on. The solid theoretical underpinning ensures that RTSP always achieves the required confidence interval. We have proposed a technique to handle large frame sizes to ensure the compliance with the C1G2 standard. We have also proposed a method to implicitly estimate the number of tags in set C. We implemented RTSP and conducted side-by-side comparisons with TH, the fastest prior tag identification protocol. Our experimental results show that RTSP always achieves the required confidence interval and significantly outperforms TH in terms of search time.

Chapter 8

RFID queries—multiple category

8.1 Introduction

8.1.1 Background and problem statement

Radio-frequency identification (RFID) has been widely used in various applications such as inventory management [67,81], localization [68], anticounterfeiting [130], and human tracking [63,129], etc. An RFID system typically consists of readers, tags, and a back-end server. A tag is a microchip with an antenna in a compact package that has limited computing power and communication ranges. RFID tags can be classified into two types: *active tags*, which use the internal battery to power their circuits, and *passive tags*, which do not have their own power source and are powered up by harvesting the energy from the reader's electromagnetic fields. The back-end server controls the RFID reader to send commands to query the tags, and the tags respond over a shared wireless medium.

In many RFID applications, tags are categorized into different categories. For example, a warehouse may categorize tags according to the brands or manufacturers of the items that the tags are attached to. We consider a set of tags where each tag has a unique ID that consists of two fields: a *category ID* that specifies the category of the tag, and a *member ID* that identifies the tag within its category. The number of categories and the number of tags in each category are unknown in advance. This chapter studies the practically important problem of *top-k queries (TKQs)*, which is to use RFID readers to query the tags so that we can quickly obtain the *top-k largest categories* and *the size of each such category*. For example, a vendor may want to know the most popular categories shipped in a day, or the least consumed types of goods in its warehouse [108].

***Multicategory RFID top-k query* problem:** given a set of tags that can be classified into ℓ categories $C_1, C_2, \ldots, C_\ell$, error thresholds $\varepsilon, \alpha \in (0, 1]$, and reliability requirements $\delta, \beta \in [0, 1)$, a multicategory RFID TKQ scheme outputs a set of k categories $\mathcal{K}$ and the size of each category in $\mathcal{K}$, which satisfy the following constraints. Let n_i be the actual size of C_i and $\hat{n}_i$ be the estimated size of C_i for $1 \leq i \leq \ell$, and $\mathcal{M} = \max\{n_j \mid C_j \notin \mathcal{K}\}$ (i.e., $\mathcal{M}$ is the size of the largest nontop-k category):

$$\textbf{\textit{Membership constraint:}} \ \forall C_i \in \mathcal{K}, Pr\left[n_i \geq (1 - \varepsilon)\mathcal{M}\right] \geq \delta$$

$$\textbf{\textit{Population constraint:}} \ \forall C_i \in \mathcal{K}, Pr\left[|\hat{n}_i - n_i| \leq \alpha n_i\right] \geq \beta \tag{8.1}$$

The membership constraint means that for any category C_i in $\mathcal{K}$, the probability that its size is larger than or equal to $(1 - \varepsilon)\mathcal{M}$ is larger than or equal to δ. The population constraint means that for any category C_i in $\mathcal{K}$, the probability that the size difference between its actual size n_i and its estimated size $\hat{n}_i$ is less than or equal to $n_i\alpha$ is larger than or equal to β.

8.1.2 Limitations of prior art

A straightforward solution is to use the tag identification protocols [77,78] to read the ID of each tag. Although perfectly accurate, they are relatively slow as they have to identify each individual tag. Another straightforward solution is to use the tag estimation protocols [70,72,73,131,132] to estimate the size of each category. Although simple, they are also relatively slow as they have to individually estimate the size of each category. The ensemble sampling (ES) protocol [67] can address our problem; however, it does not scale as its frame size is equal to the number of tags.

8.1.3 Proposed approach

In this chapter, we propose a TKQ protocol and segmented perfect hashing (SPH) for optimizing TKQ.

8.1.3.1 Top-k query

TKQ is based on the framed-slotted Aloha protocol. First, the reader broadcasts the frame size f and a random seed $\mathcal{R}$ to initialize a slotted time frame. Each tag chooses a slot $sc \in [0, f - 1]$ by calculating the hash function $sc = \mathcal{H}(C_i, \mathcal{R}) \mod f$. Thus, the tags from the same category always choose the same slot. For simplicity, we first assume that the category set $\mathscr{C} = \{C_1, C_2, \ldots C_\ell\}$ is known in advance. For each nonempty slot, if all the tags belong to the same category, then we call it a *homogeneous slot*; otherwise, we call it a *heterogeneous slot*. The reader then calculates $\mathcal{H}(C_i, \mathcal{R})$ mod f for $1 \leq i \leq \ell$ to predict whether the slot is a homogeneous slot, a heterogeneous slot, or an empty slot. In the slot chosen by a tag, the tag responds with a v-bit single-one geometric (SOG) string, denoted by $\mathbb{SOG}[0 \ldots v - 1]$, which satisfy two constraints: (i) only one bit is 1 and the other bits are 0s; (ii) the probability that $\mathbb{SOG}[j] = 1$ is $\frac{1}{2^{j+1}}$ for $j \in [0, v - 1]$. The tags transmit the SOG string using the ON-OFF keying modulation [133]: a bit 1 is represented by the presence of a carrier wave; a bit 0 is represented by the absence of a carrier wave. The combined signal $\mathbb{CS}[0 \ldots v - 1]$ is logically equal to the result of the bitwise *OR* operation on all replied strings, as illustrated in Figure 8.1.

Our intuition is that for any slot, the more tags choose it, the longer the sequence of continuous leading 1s in the combined signal is. Figure 8.2 illustrates this fact. Hence, we use the length of the continuous leading 1s observed in the combined signal to estimate the number of tags in the current slot. Consider the example in Figure 8.1, the length of continuous leading 1s in the combined signal is 2. Using the estimator given in Lemma 8.8.1, we can estimate the number of tags that choose this slot by calculating $1.2897 \times 2^2 \approx 5.16$, which is very close to the real tag number. Since the combined signal in a heterogeneous slot can only estimate the sum of

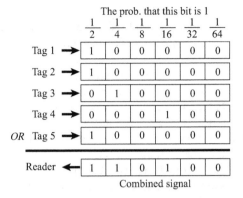

Figure 8.1 Multiple SOG strings are combined based on the bitwise OR operation

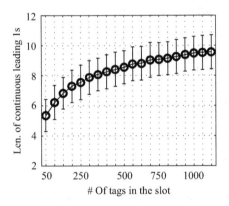

Figure 8.2 Monotonous relationship between the length of continuous leading 1s and the number of tags

the sizes of multiple categories, TKQ uses the homogeneous slots. We use multiple independent frames so that each category can occupy sufficient homogeneous slots to achieve accurate estimate. After each round of estimation, we sort the estimated category sizes to find the top-k ones.

8.1.3.2 Segmented perfect hashing

The homogeneous slots are useful for estimating category sizes. However, the ratio of homogeneous slots in the frame is up to 36.8% on average, when the frame size equals the number of categories. The low frame utilization becomes a key bottleneck of TKQ. Recall that each tag uses $\mathscr{H}(C_i, \mathscr{R})$ mod f to choose a slot. Essentially, the seed $\mathscr{R}$ determines the mapping between categories and slots. A straightforward method is to generate a series of seeds, from which we find the best seed $\mathscr{R}$ that establishes a *full bijective mapping* between all categories and the slots. However, as shown in Table 8.2, finding such a full bijective mapping seed is time consuming when the number of categories ℓ is large. For example, it takes even 9.1×10^{13} years on average when $\ell = 64$.

To improve the frame utilization of TKQ, we propose an optimization scheme called *segmented perfect hashing (SPH)*. At the beginning of each round of estimation, we put all category IDs that are under estimation into a pending set $\mathscr{P}$. Instead of finding a full bijective mapping seed, SPH just needs to find a random seed $\mathscr{R}$ that establishes a bijective mapping between m categories and the first m slots in the frame. Let $\mathscr{S}$ represent the set of these m categories. Then, the reader broadcasts $\mathscr{R}$ and f to the tags to issue a frame. After executing the first m homogeneous slots, the reader sends commands to terminate the current frame immediately. Then, we set $\mathscr{P} = \mathscr{P} - \mathscr{S}$. We repeat the above process until $\mathscr{P} = \emptyset$. Logically, a long frame is divided into multiple frame segments. The slots in each segment are homogeneous slots.

8.1.4 Technical challenges and solutions

The first key challenge is to guarantee the accuracy constraints in (8.1). We propose an unbiased estimator that leverages the average length of continuous leading 1s in the homogeneous slots to estimate the category sizes. We also calculate the variance of the estimator to measure its deviation from the real category size. To satisfy the membership constraint, we propose a fundamental theorem to recognize the larger one among two arbitrary categories with a predefined reliability δ. We determine that $C_i \in \mathscr{K}$ if C_i is larger than $\ell - k$ small categories, and $C_j \notin \mathscr{K}$ if C_j is smaller than k large categories. To satisfy the population constraint, we calculate the number of homogeneous slots required for any $C_i \in \mathscr{K}$ so that the deviation of the estimated category size is small enough to satisfy the (α, β) accuracy.

The second key challenge is to optimize the frame segment size m, which is the key factor that significantly impacts the time efficiency of SPH. Intuitively, if the frame segment size m is too large, the computation cost on the server side will be large accordingly and will dominate the total time cost. On the contrary, if the frame segment size m is too small, the current round of estimation will contain many frame segments with each requiring an extra transmission cost for transmitting parameters $\langle \mathscr{R}, f \rangle$. Essentially, the segment size m trades off between the computation cost and the communication cost. We define an *efficiency function* $\mathscr{F}_{(m, |\mathscr{P}|)}$ for measuring the time efficiency of TKQ. We calculate the optimal value of m by maximizing $\mathscr{F}_{(m, |\mathscr{P}|)}$.

8.1.5 Novelty and advantage over prior art

The key novelty of this chapter is twofold. First, we propose the *TKQ* protocol to find the top k largest categories and their sizes with guaranteed accuracy constraints. Second, we propose the *SPH* scheme to improve the average frame utilization of TKQ from 36.8% to nearly 100%. Compared with ES [67], TKQ uses a short frame size that is equal to the number of categories, whereas ES uses a frame size that is equal to the number of tags. This leads to much faster execution speed. Compared with RFID identification protocols [77], RFID estimation protocols [70], and multicategory RFID TKQ protocol ES [67], our TKQ+SPH protocol achieves 2.6–7× speedup.

The rest of this chapter is organized as follows. In Sections 8.2 and 8.3, we present the detailed design of TKQ and SPH, respectively. In Section 8.4, we review the related work. In Section 8.5, we conduct simulations to evaluate the performance of the proposed protocols. Section 8.6 concludes this chapter.

8.2   The basic protocol: TKQ

In this section, we first present the detailed design of the *TKQ* protocol. Then, we propose an unbiased estimator for estimating each category size, as well as the estimation variance. Finally, we explain how TKQ dynamically determines the top k largest categories and their sizes, meanwhile satisfying the constraints in (8.1).

TKQ includes several independent rounds of estimation. At the beginning of an arbitrary round of estimation, the reader broadcasts the parameters $\langle \mathfrak{R}, v \rangle$ to notify each tag to randomly generate a v-bit SOGs string [72,84]. Specifically, the tag first generates a v-bit random binary string $\mathbb{B}[0 \ldots v - 1]$ by returning the first v bits of the hash value $\mathscr{H}(ID, \mathfrak{R})$. Let μ represent the length of continuous leading 0s in $\mathbb{B}[0 \ldots v - 1]$. The tag sets $\mathrm{SOG}[\mu] = 1$ and $\mathrm{SOG}[i \neq \mu] = 0$. Qian *et al.* indicated that $v = 32$ is large enough [72]. Hence, we set $v = 32$ in the rest of this chapter. For example, a tag generates a 32-bit random string $\mathbb{B}[0 \ldots 31]$ as "00101...." Since the length of continuous leading 0s in $\mathbb{B}[0 \ldots 31]$ is 2, the tag sets $\mathrm{SOG}[2] = 1$ and $\mathrm{SOG}[i \neq 2] = 0$, i.e., the generated SOG string is "00100...." An extreme case is that every bit in $\mathbb{B}[0 \ldots 31]$ is 0, then, $\mathrm{SOG}[0 \ldots 31]$ will be all 0s accordingly, which does not comply with the property of a SOG string. Fortunately, the probability of this case is as small as $\frac{1}{2^{32}}$.

Then, the reader issues a slotted time frame by broadcasting the parameters $\langle \mathscr{R}, f \rangle$, where $\mathscr{R}$ is a random seed and f is the frame size. Upon receiving these parameters, each tag initializes its slot counter sc by calculating $sc = \mathscr{H}(C_i, \mathscr{R}) \bmod f$, where C_i is its category *ID*. Clearly, the tags from the same category should have the same slot counters. The reader broadcasts `QueryRep` command at the end of each slot to notify each tag to decrement its slot counter sc by one [53]. A tag will respond to the reader once its slot counter sc becomes 0. The tags from the same category C_i necessarily choose the same slot, because each tag chooses the slot by calculating $\mathscr{H}(C_i, \mathscr{R}) \bmod f$. There are three types of slots: the *homogeneous slot* in which the tags are from the same category; the *heterogeneous slot* in which the tags are from different categories; the *empty slot* that no tag chooses. We are able to predict whether a slot is homogeneous, heterogeneous, or empty, by calculating $\mathscr{H}(C_i, \mathscr{R}) \bmod f$ for $1 \leq i \leq \ell$. TKQ uses the combined signals in the homogeneous slots to estimate the corresponding category sizes. We set the frame size f to the number of categories because this setting results in the highest ratio of homogeneous slots in the frame on average [77]. The categories that pick the heterogeneous slots cannot be estimated in this round. Therefore, multiple frames with different seeds are required to let each category have a chance to occupy sufficient homogeneous slots for estimation.

Assume $\mathcal{N}$ rounds of estimation have been executed. We use a boolean variable $\theta_{i,x}$ to indicate whether a category C_i chooses a homogeneous slot in the xth round of estimation. If so, $\theta_{i,x} = 1$; otherwise, $\theta_{i,x} = 0$. We now zoom in the slot that category C_i chooses in the xth round of estimation. Let $\mathcal{L}_{i,x}$ represent the length of continuous leading 1s of the combined signal in this slot. We use s_i to denote the number of homogeneous slots that the category C_i occupies among $\mathcal{N}$ rounds of estimation, then, we have $s_i = \sum_{x=1}^{\mathcal{N}} \theta_{i,x}$. Let $\overline{\mathcal{L}}_{i,\mathcal{N}}$ represent the average length of continuous leading 1s of the combined signals in these s_i homogeneous slots. Then, we have $\overline{\mathcal{L}}_{i,\mathcal{N}} = \frac{1}{s_i} \sum_{x=1}^{\mathcal{N}} \left(\mathcal{L}_{i,x} \times \theta_{i,x} \right)$. Next, Lemma 8.1 gives an unbiased estimator that leverages the variable $\overline{\mathcal{L}}_{i,\mathcal{N}}$ to estimate the size of category C_i. Lemma 8.2 calculates the variance of the estimator. The used notations are summarized in Table 8.1.

Lemma 8.1. *Let s_i represent the number of homogeneous slots occupied by category C_i among $\mathcal{N}$ rounds of estimation, and $\overline{\mathcal{L}}_{i,\mathcal{N}}$ represent the average length of continuous leading "1s" of the combined signals in these slots. $\hat{n}_i = \omega 2^{\overline{\mathcal{L}}_{i,\mathcal{N}}}$ is an unbiased estimate of the category size n_i, where $\omega = 1.2897$.*

Proof. The detailed proof can be referred to in [72].   □

Lemma 8.2. *Let s_i represent the number of homogeneous slots occupied by category C_i among $\mathcal{N}$ rounds of estimation, and $\overline{\mathcal{L}}_{i,\mathcal{N}}$ represent the average length*

Table 8.1   Notations used in the chapter

Notation	Description
ℓ	Number of tag categories
$C_i / \mathscr{C}$	C_i is category ID, $1 \le i \le \ell$; $\mathscr{C} = \{C_1, C_2, \ldots, C_\ell\}$
$n_i / \hat{n}_i$	Size of category C_i; estimated size of category C_i
$\mathbb{SOG}[\ldots]$	Single-one geometric (SOG) string
$\mathbb{CS}[\ldots]$	Combined signal of multiple SOG strings
$\mathcal{N}$	Number of rounds of estimation repeated by TKQ
$\theta_{i,x}$	$\theta_{i,x} = 1$ when C_i chooses a homogeneous slot in the xth round of estimation; $\theta_{i,x} = 0$ for otherwise
s_i	Number of homogeneous slots occupied by C_i
$\mathcal{L}_{i,x}$	Length of continuous leading 1s in the slot that C_i chooses in the xth round of estimation
$\overline{\mathcal{L}}_{i,\mathcal{N}}$	Average length of continuous leading 1s in the s_i homogeneous slots, of category C_i, among $\mathcal{N}$ rounds
ε / α	Error thresholds
δ / β	Reliability requirements
$\Phi(\cdot) / \Phi^{-1}(\cdot)$	Cumulative distribution function of the standard normal distribution; inverse function of $\Phi(\cdot)$
$\mathscr{H}(\cdot)$	Hash function with uniform distribution
$\mathscr{R}$	Random seed
f	Frame size, i.e., the number of slots in the frame

of continuous leading "1s" of the combined signals in these slots. The variance of
$\hat{n}_i = \omega 2^{\overline{\mathscr{L}_{i,\mathscr{N}}}}$ *is* $Var(\hat{n}_i) = (\ln 2)^2 n_i^2 \rho^2 / s_i$, *where* $\rho = 1.1213$.

Proof. We represent $\hat{n}_i = \omega 2^{\overline{\mathscr{L}_{i,\mathscr{N}}}}$ by $\phi\left(\overline{\mathscr{L}_{i,\mathscr{N}}}\right)$. Then, we calculate the Taylor's
series expansion of $\phi\left(\overline{\mathscr{L}_{i,\mathscr{N}}}\right)$ around $\mathscr{E} = E\left(\overline{\mathscr{L}_{i,\mathscr{N}}}\right)$ as follows:

$$\hat{n}_i = \phi\left(\overline{\mathscr{L}_{i,\mathscr{N}}}\right) = \phi(\mathscr{E}) + \left(\overline{\mathscr{L}_{i,\mathscr{N}}} - \mathscr{E}\right) \times \left\{ \frac{\partial \phi\left(\overline{\mathscr{L}_{i,\mathscr{N}}}\right)}{\partial \overline{\mathscr{L}_{i,\mathscr{N}}}}\Big|_{\overline{\mathscr{L}_{i,\mathscr{N}}}=\mathscr{E}} \right\} \quad (8.2)$$

$$= n_i + \ln 2 \cdot n_i \cdot \left(\overline{\mathscr{L}_{i,\mathscr{N}}} - \mathscr{E}\right)$$

Then, we can calculate the variance $Var(\hat{n}_i)$ as follows:

$$\begin{aligned} Var(\hat{n}_i) &= E\left\{\hat{n}_i - E(\hat{n}_i)\right\}^2 \\ &= (\ln 2)^2 n_i^2 E\left\{\overline{\mathscr{L}_{i,\mathscr{N}}} - E\left(\overline{\mathscr{L}_{i,\mathscr{N}}}\right)\right\}^2 \qquad (8.3) \\ &= (\ln 2)^2 n_i^2 Var\left(\overline{\mathscr{L}_{i,\mathscr{N}}}\right) \end{aligned}$$

We know from [72] that $Var\left(\overline{\mathscr{L}_{i,\mathscr{N}}}\right) = \rho^2 / s_i$, where $\rho = 1.1213$. Therefore, we
have $Var(\hat{n}_i) = (\ln 2)^2 n_i^2 \rho^2 / s_i$. $\qquad \square$

Lemma 8.8.2 infers that, as the estimation is repeated round by round, the estimator variance for each category is expected to decrease gradually. After each round of estimation, we compare the estimated category sizes to dynamically determine whether a category belongs to the top-k set $\mathscr{K}$. It is nontrivial to investigate how to correctly compare two categories according to their estimated sizes. For two categories C_i and C_j, if the estimated sizes satisfy $\hat{n}_i < \hat{n}_j$, can we assert that the actual cardinality n_i is smaller than n_j? The answer is no, due to the inherent estimation variance: a small-size category may be overestimated, while a large-size category may be underestimated. Next, Theorem 8.1 tells us how to correctly compare two category sizes with a predefined reliability $\delta \in (0, 1)$.

Theorem 8.1. *Assume that two categories C_i and C_j occupy s_i and s_j homogeneous slots, respectively. Their estimated category sizes satisfy $\hat{n}_i > \hat{n}_j$; we have $Pr\{n_i \geq (1 - \varepsilon)n_j\} \geq \delta$, if the following inequality holds.*

$$\frac{\hat{n}_i - \hat{n}_j + n_j \varepsilon}{\ln 2\rho \sqrt{n_i^2/s_i + n_j^2/s_j}} \geq \Phi^{-1}(\delta) \qquad (8.4)$$

Proof. Since the estimated category sizes $\hat{n}_i$ and $\hat{n}_j$ are independent to each other, we have $E(\hat{n}_i - \hat{n}_j) = E(\hat{n}_i) - E(\hat{n}_j) = n_i - n_j$ and $Var(\hat{n}_i - \hat{n}_j) = Var(\hat{n}_i) + Var(\hat{n}_j)$.

According to the central limit theorem, we know $\mathscr{L} = \frac{(\hat{n}_i - \hat{n}_j) - (n_i - n_j)}{\sqrt{\mathrm{Var}(\hat{n}_i) + \mathrm{Var}(\hat{n}_j)}}$ asymptotically follows the standard normal distribution [11]. Then, $Pr\left\{n_i \geq (1 - \varepsilon)n_j\right\}$ can be transformed as follows:

$$Pr\left\{\mathscr{L} \leq \frac{\hat{n}_i - \hat{n}_j + n_j\varepsilon}{\sqrt{\mathrm{Var}(\hat{n}_i) + \mathrm{Var}(\hat{n}_j)}}\right\} = \Phi\left(\frac{\hat{n}_i - \hat{n}_j + n_j\varepsilon}{\sqrt{\mathrm{Var}(\hat{n}_i) + \mathrm{Var}(\hat{n}_j)}}\right) \tag{8.5}$$

To ensure the above probability is larger than the required reliability δ, we should guarantee $\frac{\hat{n}_i - \hat{n}_j + n_j\varepsilon}{\sqrt{\mathrm{Var}(\hat{n}_i) + \mathrm{Var}(\hat{n}_j)}} \geq \Phi^{-1}(\delta)$. Here, $\Phi(\,\cdot\,)$ is the cumulative distribution function of the standard normal distribution, and $\Phi^{-1}(\,\cdot\,)$ is its inverse function. Substituting the expressions of $\mathrm{Var}(\hat{n}_i)$ and $\mathrm{Var}(\hat{n}_j)$ into the above inequality, we obtain (8.4).   □

To satisfy the membership constraint, we determine that $C_i \in \mathscr{K}$ if we can find at least $\ell - k$ small categories, says C_j, such that C_i and each C_j satisfy the inequality in (8.4); $C_j \notin \mathscr{K}$ if we can find at least k large categories, says C_i, such that each C_i and C_j satisfy the inequality in (8.4). Next, Theorem 8.2 indicates how to ensure the population constraint.

Theorem 8.2. *Let s_i represent the number of homogeneous slots that category C_i occupies, $\alpha \in (0, 1)$ be the confidence interval, $\beta \in (0, 1)$ be the required reliability. To guarantee $Pr\left\{|\hat{n}_i - n_i| \leq n_i\alpha\right\} \geq \beta$, we should ensure the inequality:*

$$s_i \geq \frac{(\ln 2)^2\rho^2}{\alpha^2}\left\{\Phi^{-1}\left(\frac{1 + \beta}{2}\right)\right\}^2 \tag{8.6}$$

Proof. $Pr\left\{|\hat{n}_i - n_i| \leq n_i \cdot \alpha\right\} \geq \beta$ can be transformed into:

$$Pr\left\{\frac{-\alpha n_i}{\sqrt{\mathrm{Var}(\hat{n}_i)}} \leq \mathscr{W} \leq \frac{\alpha n_i}{\sqrt{\mathrm{Var}(\hat{n}_i)}}\right\} \geq \beta \tag{8.7}$$

here $\mathscr{W} = \frac{\hat{n}_i - E(\hat{n}_i)}{\mathrm{Var}(\hat{n}_i)}$ asymptotically follows the standard normal distribution [11]. To ensure the above inequality, we only need to guarantee $\frac{\alpha n_i}{\sqrt{\mathrm{Var}(\hat{n}_i)}} \geq \Phi^{-1}(\frac{1+\beta}{2})$. Substituting the expression $\mathrm{Var}(\hat{n}_i)$ into this inequality, we obtain (8.6).   □

Let $\mathscr{U}$ represent the set of categories that are under estimation, which is initialized as $\mathscr{C}$ at the beginning. After each round of estimation, we delete the small categories that are out of the top-k set from $\mathscr{U}$. Additionally, we remove the large categories that should be in the top-k set, meanwhile satisfying the population constraint from $\mathscr{U}$ to $\mathscr{K}$. As the estimation process goes on, the set $\mathscr{U}$ will shrink gradually while $\mathscr{K}$ grows. The estimation process of TKQ is repeated round by round until $|\mathscr{K}| = k$. Then, we obtain the k largest categories in $\mathscr{K}$ that satisfy the constraints in (8.1).

8.3   The supplementary protocol: SPH

In this section, we first explain the motivation of proposing the *SPH* protocol. Then, we use a case study to elaborate on the basic idea of SPH, which is followed by the detailed protocol design. We also provide theoretical analyses to optimize the key parameters to maximize the time efficiency of SPH. Finally, we will discuss some practical issues such as multireader deployment, and channel error, etc.

8.3.1   Motivation and challenge

The homogeneous slots are useful in the TKQ protocol. Hence, the ratio of homogeneous slots in the frame can be interpreted as the frame utilization, which is given by $\binom{\ell}{1}\left(\frac{1}{f}\right)\left(1-\frac{1}{f}\right)^{\ell-1} \approx \rho e^{-\rho}$. Here, $\rho = \ell/f$, ℓ is the number of categories, and f is the number of slots in the frame. It is easy to find that the average frame utilization is just up to 36.8% when $\rho = 1$. The low frame utilization is a major bottleneck of TKQ. Some previous work, e.g.[88], uses a bit vector to guide the tags to skip the useless slots, thereby improving the frame utilization. However, it requires the tags to be able to interpret the bit vector, such a functionality is hard to be implemented in the passive RFID tags.

   Obviously, the ideal case is that all the slots in the frame are homogenous slots, i.e., there is a bijective mapping between the tag categories and slots. Recall that, given an arbitrary seed $\mathscr{R}$, the server is able to predict the mapping between tag categories and slots even before actually executing the frame. Hence, a straightforward method is to generate a series of seeds $\mathscr{R}$, and the server tests them to choose the best one that establishes a bijective mapping between tag categories and slots. Using such a seed to issue a frame, each slot in the actual frame will be the homogeneous slot. Next, we analyze the computation complexity of this method.

   For an arbitrary seed $\mathscr{R}$, the probability that it establishes a full bijective mapping between ℓ categories and $f = \ell$ slots is $P_b = \ell!/\ell^\ell$. Therefore, we need to test $1/P_b = \ell^\ell/\ell!$ hash seeds on average, so that one of them is a full bijective mapping seed. When testing a seed $\mathscr{R}$, the server first calculates the hash function $\mathscr{H}(C_i, \mathscr{R}) \bmod f$ to map each category C_i to a slot in the virtual frame for each $i \in [1, \ell]$. Then, the reader checks whether all slots in the virtual frame are homogeneous slots. Let η represent the number of clock cycles required by the server to calculate the hash function, λ be the number of clock cycles that it takes to check the status of each slot in the virtual frame, and t_c be the duration of a clock cycle that depends on the CPU frequency of the server. Then, the average time of finding such a full bijective mapping seed should be $1/P_b \times (\ell\eta + \ell\lambda)t_c$. The numerical results in Table 8.2 reveal that the required computation cost grows sharply as the number of categories ℓ increases. For example, when $\ell = 64$, it takes even 9.1×10^{13} years on average to find a bijective random seed, using a single-threaded program. Therefore, establishing a bijective mapping between the categories and slots is a nontrivial issue. Note that, in Table 8.2, we set $\eta = 344$, $\lambda = 3$ according to [134], and the clock cycle $t_c = 4.17 \times 10^{-10}$ s with the 2.4-GHz CPU.

Table 8.2   *The computation cost of the straightforward method*

# of categories (ℓ)	# of tested seeds	Time cost on average
1	1	1.5×10^{-7} s
2	2	5.8×10^{-7} s
4	10.7	6.2×10^{-6} s
8	416.1	4.8×10^{-4} s
16	8.8×10^5	2.4 s
32	5.6×10^{12}	297.6 days
64	3.1×10^{26}	9.1×10^{13} years

8.3.2   Case study

Here, we give a case study, as exemplified in Figure 8.3, to explain the basic idea of the *SPH* protocol. Assume that there are nine tag categories $C_1 - C_9$ in the RFID system. At the beginning, a pending category set $\mathscr{P}$ is initialized as $\{C_1, C2, \ldots, C_9\}$. SPH includes several mapping processes. In the first mapping, we find a random seed $\mathscr{R}_1$ that establishes a bijective mapping between three categories (i.e., C_1, C_2, and C_4) and the first three slots. After executing the first three slots, the reader terminates the frame and sets $\mathscr{P} = \mathscr{P} - \{C_1, C_2, C_4\}$. Here, the first three slots can be interpreted as a frame segment. In the second mapping, we find another random seed $\mathscr{R}_2$ that establishes a bijective mapping between three categories (i.e., C_3, C_6, and C_7) in $\mathscr{P}$ and the first three slots. After executing the first three slots, the reader terminates the frame and sets $\mathscr{P} = \mathscr{P} - \{C_3, C_6, C_7\}$. In the third mapping, we find a random seed $\mathscr{R}_3$ that establishes a bijective mapping between $\mathscr{P} = \{C_5, C_8, C_9\}$ and three slots. After executing the last frame segment, the pending set $\mathscr{P}$ becomes $\emptyset$ and SPH terminates. Logically, we obtain a bijective mapping between nine categories and nine slots through three frame segments.

To show the advantage of SPH, we first calculate its computation cost in the example. Given an arbitrary seed $\mathscr{R}$, each category in the pending set $\mathscr{P}$ is randomly hashed to one of the $f = |\mathscr{P}|$ slots in the frame. We refer to the seed $\mathscr{R}$ as an $\langle m, |\mathscr{P}| \rangle$ *bijective seed*, if it establishes a bijective mapping between m categories and the first m slots in the frame. Let $P_{\langle m, |\mathscr{P}| \rangle}$ represent the probability that $\mathscr{R}$ is such a seed. We calculate its expression as follows:

$$P_{\langle m, |\mathscr{P}| \rangle} = \frac{\binom{|\mathscr{P}|}{m} \times m! \times (|\mathscr{P}| - m)^{|\mathscr{P}| - m}}{|\mathscr{P}|^{|\mathscr{P}|}} \tag{8.8}$$

To obtain an $\langle m, |\mathscr{P}| \rangle$ *bijective seed*, the server needs to test $1/P_{\langle m, |\mathscr{P}| \rangle}$ random seeds on average. The expected computation cost, denoted as $\mathscr{O}_{\langle m, |\mathscr{P}| \rangle}$, is given as follows:

$$\mathscr{O}_{\langle m, |\mathscr{P}| \rangle} = \frac{|\mathscr{P}| \eta t_c + m \lambda t_c}{P_{\langle m, |\mathscr{P}| \rangle}} = \frac{|\mathscr{P}|^{|\mathscr{P}|} \times \{|\mathscr{P}| \eta t_c + m \lambda t_c\}}{\binom{|\mathscr{P}|}{m} \times m! \times (|\mathscr{P}| - m)^{|\mathscr{P}| - m}}, \tag{8.9}$$

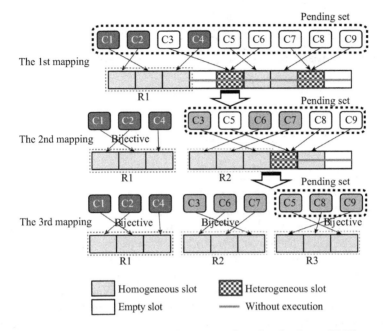

Figure 8.3 A case study of segmented perfect hashing (SPH)

where $\{|\mathscr{P}|\eta t_c + m\lambda t_c\}$ is the computation cost of testing each seed. Now, we consider the example in Figure 8.3. The overall computation cost on the server side will be $\mathscr{O}_{(3,9)} + \mathscr{O}_{(3,6)} + \mathscr{O}_{(3,3)}$. In fact, the straightforward method is a special case of SPH, where $m = \ell$. In this example, the computation cost of the straightforward method is $\mathscr{O}_{(9,9)}$. We calculate $\frac{\mathscr{O}_{(9,9)}}{\mathscr{O}_{(3,9)}+\mathscr{O}_{(3,6)}+\mathscr{O}_{(3,3)}} \approx 38.4$, which infers that SPH is about 38.4 times faster than the straightforward method in this example. Note that, SPH can show more performance improvement if larger number of tag categories are involved.

8.3.3 Detailed design of SPH

Before presenting the detailed design of SPH, we need to reclarify the relationship between TKQ and SPH as follows. Generally, TKQ includes multiple rounds of estimation. In each round of estimation of TKQ, we use SPH to improve the frame utilization. SPH includes multiple mapping processes, each corresponds to a frame segment. In SPH, each tag needs to keep a flag that has three possible values. A tag with the flag of $\mathscr{A}0$ is active and will participate in the next mapping process; a tag with the flag of $\mathscr{A}1$ will not participate in the rest of mapping processes in current round of estimation, but it will still participate in the next round of estimation; a tag with the flag of $\mathscr{B}$ is inactive and will no longer participate in the remaining rounds of estimation. Next, we will present the SPH protocol in detail, meanwhile explaining the state transition of the RFID tags shown in Figure 8.4.

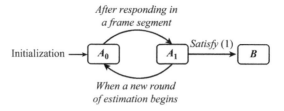

Figure 8.4   State diagram of an RFID tag in SPH

At the beginning of an arbitrary round of estimation, we use $\mathcal{U}$ to represent the set of categories that are under estimation. The reader sends commands to reset the tags whose categories are within $\mathcal{U}$ to be with the flag of $\mathcal{A}0$. The pending category set $\mathcal{P}$ is initialized as $\mathcal{U}$. The server finds an $\langle m, |\mathcal{P}|\rangle$ bijective seed $\mathcal{R}$ that establishes a bijective mapping between m categories in $\mathcal{P}$ and the first m slots in the frame. We use $\mathcal{S}$ to represent the set of these m categories, clearly, $\mathcal{S} \subseteq \mathcal{P}$. The reader uses the binary parameters $\langle \mathcal{R}, f \rangle$ to issue a frame. Each tag determines a slot by calculating $\mathrm{sc} = \mathcal{H}(\mathcal{R}, ID) \mod f$. At the end of each slot, the reader broadcasts the command QueryRep to notify each tag with the flag of $\mathcal{A}0$ to decrement its slot counter sc by one [53]. Upon finding its slot counter $\mathrm{sc} = 0$, a tag will respond to the reader with the generated SOG string and turns its flag from $\mathcal{A}0$ to $\mathcal{A}1$. Using the combined signals received in the homogeneous slots, the reader updates the estimated category sizes as well as the estimation variance based on Lemmas 8.1 and 8.2. If a category has already satisfied the constraints in (8.1), the reader will send an ACK command to turn the flag of the tags confined in this slot to $\mathcal{B}$. The tags with the flag of $\mathcal{B}$ will keep inactive and will no longer participate in the remaining rounds of estimation. After the execution of the first m slots, the reader terminates the current frame segment immediately, and the server updates the pending set by $\mathcal{P} = \mathcal{P} - \mathcal{S}$. Note that, the tags that responded in previous frame segments should have the flag $\mathcal{A}1$. Only the tags with flag $\mathcal{A}0$ will participate in the next mapping process. Similar processes are repeated until the pending set $\mathcal{P} = \emptyset$, which also means that the current round of estimation is finished.

8.3.4   Parameter optimization

The time cost of SPH contains two aspects: (i) the *computation cost* on the server side for finding the $\langle m, |\mathcal{P}|\rangle$ bijective seeds; (ii) the *communication cost* for transmitting the parameters $\langle \mathcal{R}, f \rangle$ (from reader to tags) and transmitting SOG strings (from tags to reader). The numerical results in Figure 8.5 reveal that the frame segment size m is a key factor that controls the tradeoff between the computation cost and the communication cost. The underlying reason is elaborated as follows. If the frame segment size m is too large, the computation cost on the server side will be large accordingly and will dominate the total time cost. On the contrary, if the frame segment size m is too small, the current round of estimation will contain too many

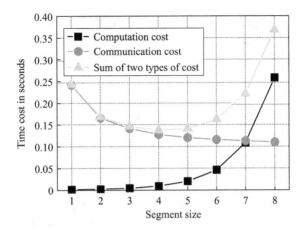

Figure 8.5   *Tradeoff between computation and communication costs.* $\ell = 50$

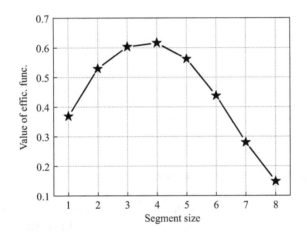

Figure 8.6   *Impact of segment size on the efficiency function.* $|\mathscr{P}|=50$

frame segments, each requiring the nonnegligible transmission cost for sending the parameters $\langle \mathscr{R}, f \rangle$. Therefore, it is important to optimize the frame segment size m to minimize the total time cost.

Given a pending category set $\mathscr{P}$, (8.8) has given the computation cost for finding an $\langle m, |\mathscr{P}| \rangle$ bijective seed on the server side. Let $\mathscr{T}_{\mathrm{act}}$ represent the actual communication cost of executing a frame segment with size m; $\mathscr{T}_{\mathrm{idl}}$ be the ideal communication cost of executing a frame segment with size m. We have $\mathscr{T}_{\mathrm{act}} = m \times \tau_u^v + \tau_d^\vartheta$ and $\mathscr{T}_{\mathrm{idl}} = m \times \tau_u^v$. Here, τ_u^v is the time to transmit the v-bit SOG strings from tags to the reader (uplink). τ_d^ϑ is the time to transmit the ϑ-bit parameters $\langle \mathscr{R}, f \rangle$ from reader to tags (downlink) for initializing the frame. We define an *efficiency function*

$\mathcal{F}_{\langle m, |\mathcal{P}|\rangle} = \frac{\mathcal{T}_{idl}}{\mathcal{T}_{act} + \mathcal{O}_{\langle m, |\mathcal{P}|\rangle}}$. Its physical meaning is that the numerator indicates the ideal time for executing a frame segment containing m slots; the denominator indicates the actual time for executing such a frame segment. Obviously, the value of $\mathcal{F}_{\langle m, |\mathcal{P}|\rangle}$ is no more than 1. In the ideal case, where it does not involve the transmission of parameters and the computation cost on the server side, the value of $\mathcal{F}_{\langle m, |\mathcal{P}|\rangle}$ equals 1. However, such an ideal case cannot be achieved because these two costs are inevitable. We calculate the expression of $\mathcal{F}_{\langle m, |\mathcal{P}|\rangle}$ as follows:

$$\mathcal{F}_{\langle m, |\mathcal{P}|\rangle} = \frac{m(m!)\tau_u^v \binom{|\mathcal{P}|}{m}(|\mathcal{P}| - m)^{|\mathcal{P}|-m}}{(m\tau_u^v + \tau_d^{\vartheta})\binom{|\mathcal{P}|}{m}m!\,(|\mathcal{P}|-m)^{|\mathcal{P}|-m} + |\mathcal{P}|^{|\mathcal{P}|}(|\mathcal{P}|\eta + m\lambda)\,t_c} \tag{8.10}$$

As exemplified in Figure 8.6, we could find an optimal value of m by maximizing $\mathcal{F}_{\langle m, |\mathcal{P}|\rangle}$. We have observed from simulations that the optimal value of m is typically less than 15. Therefore, we can quickly find its optimal value even by the exhaustive searching, which occurs offline before running our protocol.

8.3.5 Discussion on some practical issues

8.3.5.1 Identification of category IDs

For the case that the category set $\mathscr{C}$ is unknown in advance, we propose a fast approach to identify the category IDs. The reader queries the tags by broadcasting parameters $\langle R, f \rangle$, and each tag determines a slot in the frame by calculating $\mathcal{H}(C_i, \mathcal{R}) \mod f$. In the picked slot, a tag responds with its category ID C_i as well as the checksum of C_i. In a homogeneous slot, the combined signal in the category ID field will match the combined signal in the checksum field; thus, the reader is able to successfully identify the corresponding category ID. Multiple frames are repeated until all categories are identified.

8.3.5.2 Deployment of multiple readers

In large-scale application scenarios, a single reader is usually unable to cover the whole application area. Therefore, multiple readers are required to be deployed. Many excellent reader-scheduling schemes were proposed to efficiently synchronize the readers [135]. We let the query commands across all the readers be consistent, and the readers return the received data to the server. Thus, these readers cooperate like a powerful reader that can cover the whole area [70]. Then, we can migrate the proposed protocols to the multireader scenarios seamlessly.

8.4 Related work

Through comprehensive review of previous work, we summarize and classify the related work into four fields below.

Tag identification is to identify the exact tag IDs within the vicinity of the reader. There are two types of solutions: Aloha-based protocols [77] and tree-based protocols [78]. In the Aloha-based protocols, the tags content for slots in the frame to respond with their IDs. In the tree-based protocols, the reader broadcasts a 0/1 string to query the tags. A tag responds with its ID once it finds that the querying string is the prefix of its ID. A reader identifies a tag when one tag responds.

Probabilistic estimation is to estimate the cardinality of a tag set with a prede-fined accuracy constraint [59,70,72,73,131,132]. Kodialam *et al.* proposed the first set of tag estimation protocols, USE and UPE, which use the number of empty or collision slots to estimate population sizes [50]. Similarly, Zheng *et al.* proposed the PET protocol for tree-based RFID systems [131]. Qian *et al.* first proposed using geo-metric distribution hash to estimate the tag cardinality of a single set [72]. Shahzad *et al.* proposed average run-based tag estimation (ART), which uses the average run length of nonempty slots for cardinality estimation [70]. Li *et al.* proposed the max-imum likelihood estimator, which looks at the energy aspect [75]. Liu *et al.* studied the problem of RFID estimation with blocker tag [59].

Iceberg query is to identify the categories whose cardinality is above the given threshold for multicategory RFID systems. Sheng *et al.* proposed group testing scheme to rapidly eliminate the groups that contain multiple small-size categories [108]. Luo *et al.* proposed a threshold-based classification protocol, which can obtain multiple logical bitmaps from a single time frame. Each bitmap can be used to estimate the tag cardinality of a category. The categories whose sizes are obviously above the threshold can be quickly removed. The iceberg query protocols cannot be borrowed to solve the problem of TKQ, because the threshold (i.e., the size of the kth largest category) is unknown previously and hard to obtain in some applications.

TKQ is to pinpoint the k largest categories within a multicategory RFID system. Xie *et al.* made the first and only dedicated effort to address the TKQ problem [67]. In the ES protocol [67], all the tags contend for a common slotted frame, and each responds with the category ID in a random slot. ES leverages the ratio of singleton slots carrying category ID C_i to the total singleton slots to estimate the category size n_i and dynamically finds the top k largest categories. In Section 8.1.2, we have discussed the limitations of the related work.

8.5 Performance evaluation

We implemented the proposed protocols in MATLAB® on a ThinkPad X230 desktop with an Intel 2.4 GHz CPU. The simulated RFID system contains ℓ tag categories. We randomly generate the category sizes following the normal distribution Norm(μ, σ) [67]. Recall that our protocols can seamlessly work in both multireader and single reader scenarios. Same as [67,70,77], we simulate a single reader that has suffi-cient power to probe all tags. The transmission rate between the reader and tags is asymmetric. The uplink (tags to reader) rate is 53 kb/s, while the downlink rate is 26.5 kb/s. Between any two consecutive data transmissions, there is a waiting time $\tau_w = 302\mu$s [70]. The clock cycle t_c of the computer is 4.17×10^{-10} s. For clarity, the programs for finding the bijective seeds required by SPH are in a single-threaded manner. We compare TKQ+SPH with three representative protocols: the *ES* protocol [67], which is the only dedicated solution to the TKQ problem; the *enhanced dynamic framed-slotted Aloha (EDFSA)* protocol [77], which is the identification protocol specified in the C1G2 standard; the *ART* protocol [70], which is an excellent tag estimation protocol. Extensive simulations under various settings are conducted to evaluate the time efficiency and the reliability of these protocols. Each simulation result is averaged from 500 independent trials.

8.5.1   Time efficiency

8.5.1.1   Impact of k

TKQ+SPH is the fastest protocol with varying number k of concerned categories.

Default settings: Membership accuracy (ε, δ) and the population accuracy (α, β) to $(0.05, 95\%)$. The number ℓ of total categories is fixed to 100. The number k of concerned categories is fixed to 10. The cardinality of each category is randomly generated following normal distribution $\text{Norm}(\mu, \sigma)$, where $\mu = 500$ and $\sigma = 400$.

Figure 8.7(a) shows the execution time of each protocol with varying k. The execution time of EDFSA is stable because it has to exactly identify all the tags, regardless of the number of concerned categories. The performance of ART is also independent of k because it needs to estimate the population of each category in a separate manner. On the contrary, the execution time of ES, TKQ, and TKQ+SPH increases against k. The underlying reason is as follows. A larger k means that more large-size categories have to be kept in the estimation process until meeting the population constraint, which will incur more execution time. TKQ+SPH significantly outperforms the benchmark protocols, particularly when k is small. For example, TKQ+SPH runs $7\times$ faster than the fastest existing protocol when $k = 5$.

8.5.1.2   Impact of ℓ

TKQ and TKQ+SPH are the only protocols that have good scalability against the number ℓ of categories. Moreover, TKQ+SPH is the fastest protocol with different settings of ℓ. In this set of simulations, we fixed the number of concerned categories k to 10 and varied the number of all categories ℓ from 50 to 150. Each category size follows the normal distribution $\text{Norm}(500, 400)$. As the number ℓ of categories increases, the number of tackled tags also greatly grows. The numerical results in Figure 8.7(b) reveal that the execution of each protocol increases more or less with respect to the number ℓ of categories, but the execution time of TKQ and TKQ+SPH increases with the smallest rate among all protocols. The larger the number ℓ of total categories is, the better the time efficiency TKQ and TKQ+SPH achieve, as compared with other protocols. Since k is fixed, increasing the number of total categories is equivalent to increase the number of unconcerned categories. Fortunately, the unconcerned categories do not require accurate estimation and can be filtered out quickly. Therefore, TKQ and TKQ+SPH are not sensitive to the number of tag categories. We observe that TKQ+SPH is the fastest protocol with different settings of ℓ, e.g. it runs $5\times$ faster than the fastest benchmark protocol when $\ell = 150$.

8.5.1.3   Impact of μ

TKQ+SPH is the fastest protocol with different settings of μ. In this set of simulations, we fixed the number k of concerned categories to 10, and the number ℓ of total

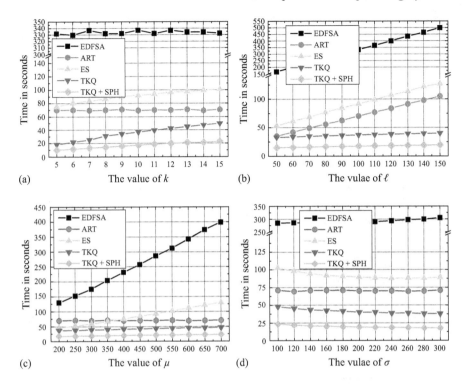

Figure 8.7   Investigating the time efficiency of protocols with different parameter settings. (a) varying k; (b) varying ℓ; (c) varying μ; and (d) varying σ

categories to 100. Each category size is randomly generated following the normal distribution Norm(μ, 150), where μ varies from 200 to 700. The simulation results in Figure 8.7(c) reveal that the execution time of ART is independent of the category size, and the execution time of TKQ and TKQ+SPH increases slightly as μ increases. The underlying reason is as follows. If we increase μ while fixing the value of σ, the relative variance of category sizes becomes small. Intuitively, it is difficult to determine which one is larger when two categories are of similar sizes. Hence, as μ increases, the execution time of TKQ and TKQ+SPH slightly grows. The execution time of EDFSA increases sharply as μ increases, because much more tags are required to be identified. The execution time of ES also increases sharply as μ increases because the used frame size should be proportional to the total tag population. We observe that ES is suitable for the RFID system that contains a large number of categories, each with a very small category size, e.g. ES runs faster than EDFSA and ART when $\mu = 200$. As μ increases, the performance of ES deteriorates. We observed that even when μ is as small as 200, TKQ+SPH still runs 2.6× faster than ES. Note that, such an improvement will be more significant when μ is larger.

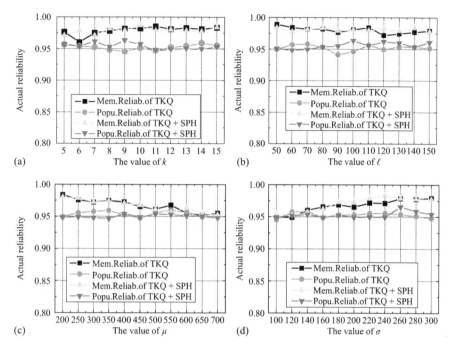

Figure 8.8   Investigating the reliability of TKQ and TKQ+SPH with different parameter settings. (a) Varying k; (b) varying ℓ; (c) varying μ; and (d) varying σ

8.5.1.4   Impact of σ

TKQ+SPH is the fastest protocol with different settings of σ. In this set of simulations, we fixed the number of concerned categories k to 10, and the number of total categories ℓ to 100. Each category size is randomly generated following the normal distribution Norm($500, \sigma$), where σ varies from 100 to 300. As illustrated in Figure 8.7(d), the execution time of ES, TKQ, and TKQ+SPH decreases as σ increases. This is because the larger σ is, the larger the variance of category sizes is. Intuitively, it is relatively easy to compare two categories with significantly different sizes. Hence, as σ increases, the execution time of ES, TKQ, and TKQ+SPH decreases. The execution time of ART is stable and regardless of the category sizes. And TKQ+SPH is about 3× faster than the fastest benchmark protocol with varying σ.

8.5.2   Reliability

Both TKQ and TKQ+SPH are able to satisfy the required reliability in various settings.

Besides the time efficiency, the actual reliability of the proposed protocols is another important performance metric. In this set of simulations, we conduct extensive simulations to evaluate the membership reliability as well as the population reliability of TKQ and TKQ+SPH. Figure 8.8(a)–(d) is plotting using different settings of k, ℓ,

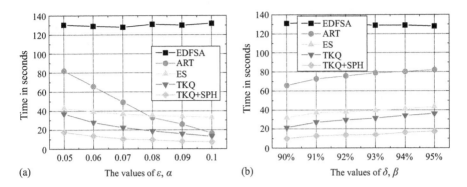

Figure 8.9   Time cost vs. query accuracy. (a) Varying ε, α; (b) varying δ, β

μ, or σ to show the impact of these key factors on the actual reliability of TKQ and TKQ+SPH. The simulation results demonstrate that the proposed protocols can meet the required membership and population constraints. Note that, the actual reliability is sometimes a bit lower than the required level. This is a normal phenomenon due to the simulation variance.

8.5.3   Time efficiency vs. accuracy

TKQ+SPH is the fastest protocol with different accuracy requirements. Figure 8.9(a) and (b) is plotted using $k = 10$, $\ell = 100$, $\mu = 200$, $\sigma = 150$. In Figure 8.9(a), we fixed the error tolerance δ and β to 95%, then varied the reliability ε and α from 0.05 to 0.1. In Figure 8.9(b), we fixed ε and α to 0.05, then varied δ and β from 90% to 95%. We observed that the execution time of EDFSA is stable regardless of the required accuracy. Except for EDFSA, the time cost of the other protocols is positively correlated to the error tolerance and is negatively correlated to the reliability. TKQ+SPH is consistently faster than all the other protocols with various settings of ε, α, δ, and β.

8.6   Conclusion

This chapter studies the problem of TKQs in multicategory RFID systems and makes three key contributions. First, we propose a *TKQ* protocol. The key idea is to use the combined signals in homogeneous slots to estimate category sizes. TKQ can quickly eliminate the sufficiently small categories, and only a limited number of large-size categories need accurate estimation. Second, we propose the *SPH* scheme to improve the frame utilization of TKQ from 36.8% to nearly 100%. We theoretically maximize the time efficiency of the proposed protocols. Third, we conduct extensive simulations to evaluate the protocol performance. The simulation results show that TKQ+SPH achieves not only the required accuracy constraints but also a 2.6–7× speedup over prior protocols.

Chapter 9
RFID privacy and authentication protocols

9.1 Introduction

Radio-frequency identification (RFID) tags are cheap, simple devices that can store unique identification information and perform simple computation to keep better inventory of packages. This feature provides a significant advantage over barcodes, allowing them to be used in applications throughout various fields such as inventory tracking, supply-chain management, theft-prevention, and the like. However, unlike barcodes, these tags have a longer range in which they are allowed to be scanned, subjecting them to unauthorized scanning by malicious readers and to various attacks, including cloning. Therefore, a security protocol for RFID tags is needed to ensure privacy and authentication between each tag and their reader. This chapter provides a general look over various security approaches created in recent years. These approaches include separate devices that were developed to protect an RFID tag and low-computation algorithmic protocols developed within the tag itself, two of which were developed by the authors of this chapter. The chapter is concluded by discussing the future direction of RFID security and some open research issues concerning its field of study.

RFID tags are small electronic components that are used to identify and track objects. They have applications in various fields such as inventory tracking, supply-chain management, theft prevention, and the like. An RFID system consists of an RFID tag (i.e., transponder), an RFID reader (i.e., transceiver), and a back-end database. An RFID reader consists of an RF transmitter and receiver, a control unit, and a memory unit. These instruments work together to transfer and receive information stored on radio waves between the reader and an antenna attached to an RFID tag. This information interacts with stored items upon a back-end database that some readers are able to connect to. Depending on the type of the tag, they too have the capability to perform different functions with the information transferred from a reader.

There are three broad categories of RFID tags: passive, semipassive, and active. Passive tags are powered by the signal of an interrogating reader and can only work within short ranges (a few meters). Active tags maintain their internal state and power transmission using a battery. Semipassive tags are battery assisted tags that use some battery power to maintain their internal volatile memory but may still rely on the reader's signal to power their transmission. They can initiate communication and operate over longer ranges (several meters) but are also more expensive and bulkier than passive tags. Passive tags, however, are also more popular and cheaper, making them more likely to be used within the broad range of applications stated earlier. Therefore, this chapter will only focus upon devices and protocols that have been designed for passive tags.

RFID tags are able to uniquely identify individual items of a product type, unlike barcodes, which only identify each product type. This is particularly useful when the transaction history of each item needs to be maintained or when individual items need to be tracked. Furthermore, RFID tags do not require line-of-sight reading like barcodes, increasing the scanning process of a tag significantly. Due to these and other advantages that RFID tags have over barcodes, RFID is increasingly becoming more popular and is expected to replace the current barcode technology in the near future. However, there is also a growing concern among people about consumer privacy protection and other security loopholes that make RFID tags an easy target for malicious attacks. Passive RFID tags in their current form are vulnerable to various types of attacks, and thus, there is a pressing need to make this technology more secure before it is viable for mass deployment. Therefore, privacy and authentication are the two main security issues that need to be addressed for the RFID technology.

The two primary concerns of privacy with RFID tags are clandestine tracking and inventorying [141]. Clandestine tracking deals with the issue of a nearby RFID reader being able to scan any RFID tag, since these tags respond to readers without discretion. Clandestine inventorying on the other hand is a method of gathering sensitive information from the tags, thus gaining knowledge about an organization's inventory. An organization called EPCGlobal [140] manages the development of the Electronic Product Code (EPC), a code in RFID tags that is equivalent to the code used to store information in a barcode. EPC compliant RFID tags have fields to store the manufacturer code and the product code that makes it easy to follow the inventory patterns of a store [141] or the assignment of ID numbers to employees of a business, for example.

RFID privacy is already a concern in several areas of everyday life. Here are a few examples. Automated toll-payment transponders, small plaques positioned in windshield corners, are commonplace worldwide. In a recent judiciary, a court subpoenaed the data gathered from such a transponder for use in a divorce case, under-cutting the alibi of the defendant [150]. Some libraries have even implemented RFID systems to facilitate book checkout and inventory control and to reduce repetitive stress injuries in librarians. Concerns about monitoring of book selections, stimu-lated in part by the USA Patriot Act, have fueled privacy concerns around RFID [151]. Lastly, an international organization known as the International Civil Aviation Orga-nization has promulgated guidelines for RFID-enabled passports and other travel documents [143,152]. The United States has mandated the adoption of these stan-dards by 27 "visa waiver" countries as a condition of entry for their citizens. The mandate has seen delays due to its technical challenges and changes in its technical parameters, partly in response to lobbying by privacy advocates. One may see how verification of the information stored upon the passport would also become an issue as well. This brings us to the other security threat in RFID authentication.

Authentication is another major security issue for RFID tags. Privacy deals with authentic tags being tampered by attacking readers, while authentication deals with valid readers being misled by deceptive tags. One example where authentication would play a useful role is when scanning counterfeit tags. It has been shown that one can rewrite what a tag emits onto another tag, effectively making a clone [141]. Therefore, authentication is as much of a concern as privacy is.

The key challenge in providing security mechanisms to passive RFID tags is that such tags have extremely weak computational power because they are designed to be ubiquitous low-cost (i.e., a few cents) devices [156]. Numerous solutions have been developed to solve both security threats for RFID tags. These solutions include separate devices that were developed to protect an RFID tag and low-computation algorithmic protocols developed within the tag itself. This chapter discusses the advantages and disadvantages of a few of these solutions from which the field of RFID security has contributed. Insight is then provided upon the current state of RFID security and its future direction.

9.2   Premier RFID authentication and privacy protocols

Upon realizing the need to provide consumer privacy to RFID tags, researchers first began to disable the tag upon scanning. In other words, once a product has been through the checkout procedure, a magnet-like device would disrupt the wiring of a tag, disabling it from use again. This procedure is currently used in many bookstores and other retail environments to prevent theft. However, this approach is not suitable for all items. Therefore, researchers began borrowing the concepts of well-known cryptography methods to not only establish privacy but provide mutual authentication for tags that contain a higher security risk such as identification badges and electronic passports. This section describes the disadvantages of disabling a tag and looks into how cryptography methods will compare to RFID tags.

9.2.1   Tag "killing" protocols

The first approach to dealing with consumer privacy was developed by the company that will oversee the barcode to RFID transfer, EPCGlobal Inc. Their approach is to just "kill" the tag [140]. In other words, the tag will be made inoperable, allowing it not to be scanned by malicious readers. This process is done by the reader sending a special "kill" command to the tag (including a short 8-bit password). For example, after you roll your supermarket cart through an automated checkout kiosk and pay the resulting total, all of the associated RFID tags will be killed on the spot.

Though killing a tag may deal with consumer privacy, it eliminates all of the postpurchase benefits for the consumer. One example of these types of postpurchase benefits is items being able to interact with what are being called "smart" machines. For example, some refrigerators in the future will interact with the RFID tags on food items. This will allow the refrigerator to scan what items you normally buy, and once it notices that so many items have been removed over a period of time, it will inform of what items are missing so you may purchase some more. Another example of a "smart" machine would be a microwave. The microwave would scan the RFID tag from the purchased item and automatically set the timer to the correct amount of time needed. From these examples, you can see that killing a tag would not be an appropriate approach to deal with consumer privacy.

9.2.2 Cryptography protocols

Due to the invention of "smart" machines and other devices that will need to reuse a tag's information repeatedly, researchers began developing ideas that will insure one's privacy and mutually authenticate a tag and a reader to one another without actually disabling the tag upon its scanning. One of the first set of approaches developed to fit this criterion borrowed the concept of the public key cryptography approach [142,153–155]. The cryptography approach is a simple and well-defined algorithm to be implemented. An example of this type of protocol would be as follows. Consider a matching pair of encryption public and private keys for both a reader R and a tag T. Each tag would initially be embedded with its reader's public key, R_{pu}, and its own unique private key, T_{pr}. Upon query from a reader, both a tag and the reader themselves may mutually authenticate each other using the tag's embedded keys. Tag T would encrypt an arbitrary nonce n with its private key within an additional encryption of the reader's public key $R_{pu}(T_{pr}(n))$ and send it to the reader. Once the reader has decoded the outer encryption using its private key R_{pr}, it will search its back-end database and obtain the tag's matching public key T_{pu}, decrypting the inner lock to retrieve the nonce from the tag. The reader would then reencrypt this nonce with its own private key, and similar to the tag, it would establish a second encryption around the previous one with the tag's public key, $T_{pu}(R_{pr}(n))$. Before establishing this encryption, the reader would use its keys in combination with the current tag's to form a temporary key T_k, which would be sent along with the nonce to the tag. Upon retrieval and decryption of the message sent from the reader, the tag would use this T_k to send any further information between itself and the reader, such as a person's job title, to enable them access to a certain location of a building. Figure 9.1 illustrates such a protocol. For a more time-based protocol, any important information from the tag would be embedded within the first message sent, such as a product's EPC [140] for a supply-chain marketing system. This approach prevents the unwanted listening of

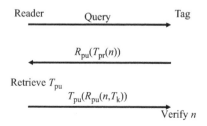

Figure 9.1   An example of cryptography protocol

the transmitted messages from a tag unless a hacker was to retrieve the private key of that tag, which is very unlikely in a given short period of time.

Despite the security strength of this approach, public key cryptography (especially for execution within RFID tags) requires strong computational power in order to establish encryption upon the transferring messages. This not only increases the size of the tag to become quite large, but it also significantly increases the cost of each tag, causing this approach to be used within very limited sectors of RFID security.

9.3   RFID privacy devices

As explained earlier, the elongated broadcasting range of an RFID tag is susceptible to unwanted scanning (or reading) by a hacker or any malicious third party. This fact is more prevalent when dealing with the issue of privacy; whether it be consumer privacy (gathering information about previously purchased items) or something of a more secure nature (retrieving job or governmental information from a identification badge). To combat this issue, many of the earlier attempts suggested carrying or attaching an external device to protect a tag from unwanted access. The following three sections provide examples of such approaches.

9.3.1   Faraday's cage

One of the first approaches in dealing with consumer privacy involves what is called a Faraday cage [141]. A Faraday cage is container made of metal mesh or foil that is designed to block certain radio frequencies. In fact, this method of shielding an item with certain metal material is known to be used by thieves when trying to surpass shoplifting detection systems. Recently, the US State Department has even indicated that US passport covers will include metallic material to limit RF penetration, thus preventing long-range scanning of closed passports [143,152]. This approach, however, does come with its disadvantages. The main disadvantage is that the cage is not designed to fit around certain items, such as wristwatches, containers, and bigger items such as televisions or computers. This disadvantage limits the use of this approach, restricting it from more commercial investments such as the supply-chain market.

9.3.2   Active jamming device

Another approach dealing with consumer privacy is called the active jamming approach [141]. This approach will allow an individual to carry a device that would block nearby RFID readers by transmitting or broadcasting its own signals. However, this approach could be illegal if the broadcast signaling power of this device is too high. This might cause the jammer to interfere with surrounding legitimate RFID readers where privacy is not a concern, disrupting a company's business. Therefore, due to the legality of this approach, it is not a suitable solution for the privacy protection of RFID.

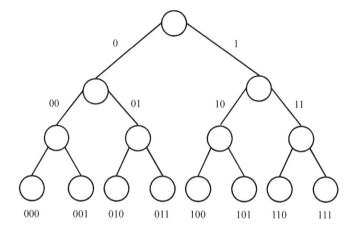

Figure 9.2   A singulation protocol tree

9.3.3   Blocker tag

There have been multiple studies related to RFID security in the past few years, and enumerating all of them is beyond the scope of this chapter. However, Juels has discussed a number of these techniques in detail and highlights the pros and cons of each method in [141]. In particular, we would like to discuss a technique of his that one of our protocols was inspired from, the privacy bit concept from [144].

The approach Juels uses in [144] is similar to that of the "jamming" approach described earlier; however, its effect is not as strong upon operation, cleverly interacting with the RFID "singulation" protocol to disrupt only certain operations. The singulation protocol for RFID tags is a tree-based method upon which a reader may tell apart multiple tags scanned at the same time. This process works by repeatedly querying all present tags within the area, distinguishing each separate tag by keeping a count of the number of collisions a reader has received. For a more detailed description of this procedure, consider the example given in Figure 9.2. This figure represents a tree with a depth of 3 containing $2^3 = 8$ tag serial numbers represented at each of its leaves. Assume, we want to distinguish between tags "001" and "011." A reader would first query for all tags with "0" prefix. This query would return a collision, since both tags have this criterion. Upon querying for a "1" prefix, no signals would be returned; therefore, the reader would not continue to query for any tags beginning with that prefix. The singulation protocol would then follow the recursion upon the collision points of the tree as before to eventually reach the corresponding leaves "001" and "011," receiving only one response back denoting their presence.

In order to interact with this protocol, in [144], Juels uses a special bit in tags called a "privacy bit" that can take a value of 0 or 1 and can be easily toggled by a reader after authenticating with a unique pin for that tag. The tag bears a value of 0 when inside a store, indicating that it has public access and on checkout the tag is moved to a "private" zone by flipping the privacy bit to 1. But doing just this does not ensure security of the tag. An additional specialized tag called the "blocker" tag must accompany the tag in question in order to secure it [144]. The blocker tag confuses malicious readers into thinking that tags with all possible values are present with the bearer. The tag specifically achieves this by sending the corresponding value a reader is currently querying for within the singulation protocol. For example, if a reader is asking for a serial number beginning with "11," the blocker tag would send this value in order to collide with any tags that may actually contain this value, confusing the reader upon reaching the leaves of its tree. This "confusing act" can either be done to overwhelm the reader (full blocker) or in a "soft" or polite way (partial or selective blocker) [145]. A soft blocker, however, would only interact with leaves that are within a "privacy zone." For example, as described earlier, the privacy bit of a tag would change from 0 to 1 upon authentication of tag. The soft blocker would then only protect tags with a prefix of "1," allowing other items within the area to be scanned if necessary. Either way, the tag is secure only in the presence of the blocker tag.

This concept is specifically designed to promote consumer privacy; however, given the difference in signaling strength of each tag, they may hold a better place within different retail environments. A normal blocker tag may be used within cellular phones to disrupt malicious transmissions trying to attack or retain information from different calls or text messaging. A soft blocker would be more suited for the supply-chain market, and thus would be embedded within a grocery or shopping bag. This temporary method of security provides the advantage of allowing "smart" machines as described earlier access to certain RFID-related items without any additional processing. However, as the previous two devices, this concept has its flaws as well. Even a well-positioned blocker tag has a chance of failing given the unreliable transmission of RFID tags [141]. Also, readers may eventually evolve in exploiting weaknesses to blocker tags and overtake their signal strength [160]. In order to fully understand the attacks and defenses upon this approach, research and evaluation will have to continue before any deployment is considered.

9.4 RFID protocols based on hash functions

Upon the failure of separate mechanisms devised to provide efficient authentication to RFID tags, researchers began developing schemes to provide such security within the tag themselves. A popular method among many protocols in achieving such security involved using an encryption-like method to secretly transfer messages between a reader and a tag, known as hash functions. A hash function is any mathematical function or well-defined method that rearranges any given data into a reasonably small integer, normally providing use as an index for an array [159]. Hash function algorithms such as SHA-1 [157] and MD5 [158] have been widely accepted as a secure form of protection for transferring data within a limited computational range. This section explores one of the original hash-based approaches developed and some developed upon the improvement of the searching time for a tag within a database.

9.4.1   Hash lock: the original hash function-based approach

One of the first protocols, which were later developed, is known as *hash lock* [139]. This hash function-based approach deals with unlocking a tag value through hash-based results to gather its secret information. The tag begins by starting in a "locked" state, where a reader sends a lock value to the tag, $lock = hash(key)$, where the key is a random value. This value is stored within the tag's reserved memory location (i.e., a Meta-ID value), wherefore the tag enters the locked state, not allowing any information about the tag other than what will need to be given during the authentication process. To unlock the tag, the reader must send the tag the original key that was used to make its Meta-ID value. Upon receiving this value, the tag performs a hash function against the key and compares it to its Meta-ID value. If it matches, the tag unlocks itself, allowing its EPC [140] to be responded to readers upon forthcoming cycles of queries. This protocol is very simple and straightforward in providing security against the breach of secret information within the tag, i.e., its EPC. Since authorized readers only know the original key value of each tag, only they are able to unlock the tag for its information, upon which they lock the tag again after reading the code.

Despite its simplicity, this protocol provides a heavy breach in security. It fails to provide mutual authentication between the tag and reader, only authenticating the reader since it must provide the key value that should be unknown to the public. This breach can simply be exploited by a malicious third party scanning the tag for its Meta-ID value. This value in turn would be randomly broadcasted to nearby readers where eventually the key value for that particular tag is returned. The party may then use this information to send to the original tag, obtaining its EPC and other sensitive information.

To combat the above-mentioned flaws, the same authors of the chapter [139] devised a new approach, dealing with randomization. The emphasis here was to disguise the Meta-ID value with a random number each query, thus the tag and its value could not easily be traced. Therefore, an additional pseudorandom number generator is embedded into the tag for this approach. Similar to the original hash-lock protocol, a tag will prestore a key value (known as an ID) to place itself in a locked state. However, this approach will not store the hashed result of this ID, but rather upon each query, the tag will hash its ID with a random value given by its pseudorandom generator, resulting in $hash(ID_k, r)$, where k represents the kth tag among a number of tags within the system, $ID_1, ID_2, \ldots, ID_k, \ldots, ID_n$.

Upon the reader's query of a tag, the reader will obtain two values. These values included the random number generated by the tag and the hash function result value generated by the hash against the ID value of the tag and its current random number at that time. To unlock the tag, a reader must send the tag its original ID value. Therefore, the reader will begin to search its back-end database containing all ID values for all tags, where it must repeatedly perform a hash function against each separate value with the random number given from the tag. This will allow the reader to compare each hashed result against the one sent from the tag, wherein if they match, the reader will obtain the matching kth ID value, returning it to the tag to unlock itself. Once the tag is in an unlocked state, any reader may perform a query to gain the tag's EPC information.

In addition to successfully achieving security on RFID tags, this approach also provides location privacy. In the previously developed protocol Hash Lock, each tag still reveals its Meta-ID. However, this approach only discloses a random number and a hashed value based from that number. Therefore, a malicious third party cannot trace a specific tag (i.e., a product of a store) based from its Meta-ID. In this case, the randomized hash-lock protocol is able to provide location privacy.

Despite the vast improvements that this randomized protocol presents over the original hash-lock scheme, this approach may not be suitable for all cases. Due to the fact that the reader must search through as many ID values as possible to find the matching hash result, its running time is that of $O(n)$, for n tags within the system. Therefore, this approach would have a vast scalability problem when the number of tags for a system increases enormously. The additional cost of producing a pseudorandom generator for a tag also presents another problem for a system of that scale.

9.4.2 Tree-based approaches

As explained earlier, the randomized hash-based approach in [139] fails to maintain a less than optimal running time and low-cost tags against a system whose number of tags increase expeditiously. However, as this type of system's popularity continues to expand overtime, the search to find an efficient algorithm grows larger. In an effort to decrease the running time of a hash-based function while maintaining its security, authors of [136–138] developed an approach which improves the key search efficiency from linear complexity to logarithmic complexity. The key in achieving this is based upon the structure of the back-end database, which is set up as a tree-based graph. Every node will then require $O(\log N)$ search time, as proven in multiple mathematical theories. One of the first groups to use a tree-based approach was Molar *et al.* [138], who in their approach used a challenge–response protocol. This procedure required multiple rounds to identify a tag, each round consisting of three messages between both parties. Due to the tree-based nature of the back-end database, however, this caused each round to have an $O(\log N)$ running time, incurring relatively large communication overhead between each message. Therefore, another chapter [136] made an improvement upon this algorithm that shortened the length of a round to only one message from the tag, providing no further interaction between the tag and the reader. To further explain the concept of tree-based protocol, a detailed explanation of the procedure in [136] is provided below.

The back-end database for a reader in a tree-based approach consists of a set of keys formed in a binary tree manner. Consider the tree in Figure 9.3. Each node contains a distinct key, and each tag is denoted to each leaf node. Therefore, there exists a unique path of keys from the root to each leaf node, whereas these set of keys are assigned to each leaf node, which is used for authentication. For example, tag T_8 contains keys k_0, $k_{1,2}$, $k_{2,4}$, $k_{3,8}$. When the reader R authenticates T_8, it first sends a nonce n to the tag. The tag T_8 then encrypts its set of keys with n by executing a hash on both values and sends the result back to the reader. The reader then searches for the leaf node corresponding to the tag by repeatedly hashing the sent nonce n with

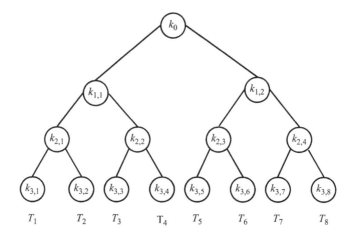

Figure 9.3   A static binary tree with eight tags

its binary tree of keys and matching those results to the sent list of encrypted results from the reader. If such a path exists to a leaf node, T_8 is identified and reader R regards this tag as valid.

From the above procedure, we can tell that the path to a tag will end up sharing certain keys with other similar paths to different tags. For example, tags T_8 and T_7 share key $k_{2,4}$, and of course, all tags will share the root key, k_0. An advantage in using this static architecture is that its running time is logarithmic. For example, in Figure 9.3, any identification of a tag only needs $\log_2(8) = 3$ search steps. This structure, however, also presents a security flaw in that if a tag is compromised, an adversary would obtain multiple paths from the root node to a leaf node, including the keys on those paths. Since the keys of the architecture are never updated, the captured keys will still be used by the uncompromised tags, allowing further knowledge of what key combinations an uncompromised tag may contain.

A practical solution to solving the above-mentioned problem is to update the keys after every authentication; however, the aforementioned static tree architecture cannot handle such a task due to its complexity. For example, if one were to update tag T_1, they would have to change keys k_0, $k_{1,1}$, $k_{2,1}$, and $k_{3,1}$ partially or totally. This in turn would single-handedly affect every path of tree, not allowing the other nonauthenticated tags to update themselves in order to authenticate correctly with a reader. Therefore, using this architecture, every tag would need to be updated periodically and simultaneously (along with static tree used by the reader) to continue the synchronization of the approach. Unfortunately, this solution is not practical in large scale systems with hundreds or millions of tags. An alternative solution would be to collect only those tags effect by each authentication of another tag's path periodically; however, this idea would be more cumbersome than the first in trying to collect a number of tags only affected by one tag changing its keys. Therefore, we have developed our own protocol called "HashTree," a dynamic key-updating algorithm for private authentication in RFID systems. Our protocol is explained in the next subsection.

9.4.3 HashTree: a dynamic key-updating approach

Though tree-based approaches have an efficient search time, they lack in the long-term security guarantee. Due to the infrastructure of a tree-based approach, sets of keys in the path between two or more tags are commonly shared. Consequently, if one tag compromised, this could lead to the leaking of information about other tags within the system. In order to resolve this issue, the keys of both the reader and the tag must be simultaneously updated. To our knowledge, [146] is the first chapter to discover this flaw and develop its own dynamic key-updating approach to this problem. In [146], the back-end database for a tree contains a set of keys k in a similar binary tree structure from that of [136]. In order to update the tree, however, each node contains an additional temporary key tk. Initially, each temporary key equals the value of the current key for each node, $tk = k$. The authentication procedure for a tag is essentially the same as a normal tree-based approach, where the tree is repeatedly searched for a path leading to a matching leaf node that represents a tag. Upon authentication of a tag, the reader performs a hash function upon k and sets its results as its current key value, $k = h(k)$. The temporary key is then always updated to the current key's value beforehand. This is to insure that every search for a key's path to its leaf node is correctly calculated, since the protocol will check both the current key value k and the temporary tk value for a node during authentication of a tag. Though this protocol prevents many of the problems with [136], it is only suitable for systems that plan on never scanning the same tag twice. This is due to the value of the temporary key only being capable of holding one key value at a time. If a tag were to be scanned multiple times, the temporary tk value of a node would only hold the ith $- 1$ current key value of a node of i scans. This would not only allow paths sharing nodes with this tag's path to be wrongfully unauthorized but would eventually misauthorize the repeatedly scanning tag as well. Therefore, we have developed the "HashTree" protocol that provides a simple dynamic key-updating system for tree-based approaches. This approach simultaneously updates the keys of a reader and tag upon each authentication, while still allowing nonpreviously queried tags to continually authenticate themselves with ease.

Our HashTree protocol is comprised of three components: *system initialization, tag identification*, and *system maintenance*. The first two components are very similar to the static tree-based approaches presented in [136] and perform the basic identification functions. However, unlike the previously described dynamic key approach [146], our protocol secures the RFID system from the *compromising attack* without actually updating a reader's back-end database. Lastly, the third component is used to direct the joining and disjoining of tags to and from the system.

Our protocol begins by providing a similar arrangement for a reader and a tag of that from [136] chapter. The back-end database of a reader contains the similar binary tree like structure that both [136] and [146] provide. Assuming that there are N tags T_i (where $1 \leq i \leq N$) and a reader R in the RFID system, reader R will assign the

N tags to N leaf nodes of the balanced binary tree S. Each node in the tree S will be assigned a preset key k. Upon the introduction of a tag T_i, the reader will distribute a path of k keys to a tag denoting an unassigned leaf node within the tree. However, unlike previous protocols, these keys will not be stored in their natural state within the tag. Instead, an array of hashed results from a randomly generated nonce n by the reader against each k key from the original given set of keys will be stored in tag T_i along with the n value used to generate those results. Storing the key values, this way is very important to the logic of the algorithm, which will be explained shortly.

As stated earlier, our authentication process for a tag is very similar to that of the protocol from [136]. However, the key difference is in how each tag has a precalculated set of hash results instead of storing the exact keys from the tree in a reader's back-end database. Upon query from a reader, the tag will send its array of hash results including the nonce n used to calculate them as described in the earlier paragraph. The reader will then take the n value and hash it against every k value of a node starting from the root node in a repetitive logarithmic manner until it has reached the matching leaf node corresponding to that tag. Assuming that a path of nodes containing a matching hash result for the array of results sent is traversed and found, the tag is then authenticated. After authentication of the tag, each tag will be updated to prevent the *compromising attack* described earlier. To achieve this, the reader will generate another random nonce n_2 different from the one given by the tag. This nonce will then be used by the reader to perform another set of hash functions in accordance with the tag's T_i keys as recorded in the reader key tree. For a clearer view of what's being transmitted, refer to tag T_3 in Figure 9.3. Using the nonce value n_2, the sequence of hash function results against the original set of keys for tag T_3 would be $(h(k_0, n_2), h(k_{1,1}, n_2), h(k_{2,2}, n_2), h(k_{3,3}, n_2))$. This new list of hash function results and the nonce n_2 generated will then be sent to the tag, thereby updating its key values.

Before explaining the system maintenance portion of this section, let's take a moment and explain the theory behind only updating the tag's values and not the reader's. The key behind chapter [146] is that it assumes an attacker may gain access to the secret values of a tag, therefore compromising a reader's back-end database key tree system. This is due to each tag's set of keys having a direct correlation with another tag's set of keys, as earlier described within the description of the static tree chapter [136] in Section 9.4.2. Our algorithm handles this type of compromise attack in several ways. First, by updating a tag's set of key values, it can no longer be traced to other tags held within a secure holding area. For example, if this protocol were used for supply-chain management within a retail store, all tags bought from the customer could not be traced to items held within the store. More importantly, the process in which we assign the tag initially and after authentication (i.e., using hash results instead of original key data) blocks the attacker from knowing the actual key values held within the key tree of the back-end database connected to a reader. Because of this, key locations from one tag to another (even tags that originally share a key path) are not associated with one another. This is one of the main reasons we do not have to update the key tree of the reader. To explain this further, consider tag's T_1 and T_2 in Figure 9.3. Both tags in this tree structure share the key $k_{2,1}$. In a static tree, compromising one of these tags would compromise the secret of the other since each

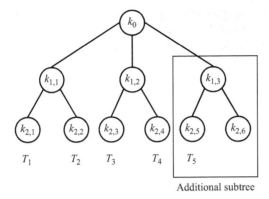

Figure 9.4 *Example of an additional subtree*

key is given to the tag in its normal form. However, our algorithm produces a different hash result of this value for both tags since each tag would have a different nonce (i.e., $h(n_1, k_{2,1}) \neq h(n_2, k_{2,1})$); therefore, the attacker would have no way of relating these two values together.

Lastly, in a real-time execution of this protocol, users may need to simultaneously remove and add additional tags to the key tree; therefore, we provide a series of small steps to perform such tasks. Assume an additional tag T_i needed to be added to the system. This task may be performed in one of two ways. The reader will first search through its list of leaf nodes; if an empty space is available, the tag's information will be placed there. However, if all the leaf nodes are currently pointing to a tag, a new subtree will have to be established. This subtree will be of length d corresponding to the depth of the tree that currently exists in the back-end database. Each node of this subtree will be assigned a randomly generated key, and the tag T_i will be assigned these values with a random nonce as described in the *system initialization* portion of the protocol as before. An example of the addition process where an empty node could not be located is shown in Figure 9.4.

Consequently, one may need to withdraw tags from the system as well. This simple task is done by emptying the leaf node associated to tag T_i. This allows for an additional tag to be associated with that node as described in the previous paragraph.

A secure authentication protocol should meet the following security requirements [136]: *privacy, untraceability, cloning resistance, forward security*, and *compromising resistance*. Specifically, the untraceability requirement deals with the notion that a tag's output should be able to correlate itself with a tag; otherwise, this tag may be traced by attackers. From looking at our protocol, some may argue that even though each tag contains a different nonce n value, the output is the same upon multiple queries. However, the updating of each tag's values upon authentication prevents this attack from occurring. As earlier stated, this protocol is made general enough for a system to interact with tag such as an identification badge, for example. Therefore, assuming this badge is used everyday to access a building, it would be difficult for an attacker to keep track of this type of tag unless the person possessing it is being physically followed throughout the day. Another issue one may bring is that of privacy. There is no outlining protection upon an attacker gaining the current values from the tag and performing a replay attack. If this presents itself as an issue for a system, then each tag may simply use a Faraday cage as explained in Section 9.3 to protect it until its ready for use, similar to many new electronic passports providing the same covering for extra security [143,152]. With this modification, our protocol successfully fulfills each security requirement for a tag, thus making it cost-effective and secure enough for deployment within a real-time RFID passive tag authentication system.

9.5   Other RFID authentication and privacy protocols

While many protocols were developed from earlier approaches such as most tree-based approaches (including our "HashTree" protocol), other researchers decided to take another route. The next protocol you will read about is developed by Juels [147] and involves the enablement of a passive tag to relabel itself, like many of its active counterparts. The idea behind this protocol combined with Juels' earlier soft-blocking [145] protocol makes another one of our protocols and the last one of the chapters, "RFIDGuard: An Authentication and Privacy Protocol Designed for Passive RFID Tags."

9.5.1   Minimalist cryptography

Many active tags contain the ability to relabel themselves in a fashion that is indistinguishable to third-party malicious attackers but may still be authorized by certified readers. Since passive tags do not contain enough power to withhold this ability, many researchers began developing additional items to somehow block or distort the transmission of the tag from hackers, as described in Section 9.3. However, Juels has developed a protocol called the "minimalist" system [147], allowing the relabeling of a passive tag within its limited computational capabilities. Within this system, each tag contains a small set of pseudonyms; a different pseudonym is then given to a reader upon query since each tag will rotate through its list of pseudonyms per scan. The security holds in that only an authorized reader will contain the entire set of pseudonyms. An unauthorized reader does not contain each pseudonym for a tag and thus will not be able to gain any secure information from the tag given its different appearances. A more thorough example of this scheme is provided below.

As previously stated, each tag contains an array of pseudonyms α_i, where i denotes the current pseudonym of a tag for m pseudonyms, $1 \le i \le m$. However, the protocol would not be secure from these pseudonyms alone. If that were the case, the minimalist system would be vulnerable to the cloning attack, an attack of which many static designed protocols for RFID tags suffer from [148]. In this case, an attacker would query the tag, thus obtaining its current pseudonym α_i and replay this value to the reader, allowing itself to be recognized as the currently scanned tag. To combat this issue, a tag only verifies itself to a reader after the reader has authenticated itself to the tag. To accomplish this, for every pseudonym α_i, each tag contains two additional key values, β_i and γ_i, for which the tag and reader will authenticate themselves with. Upon query by a reader, a tag will respond with its current α_i pseudonym value. Assuming the value is valid, the reader will retrieve the tag's corresponding β_i and γ_i values from its back-end database and send the tag β_i. Upon authentication of the reader, the tag will respond with its corresponding γ_i value. As you can see, this protocol mimics that of a simple challenge–response protocol, but one that is designed upon pseudonym rotation.

In order to successfully achieve long-term security for a tag, one must continually update its α_i, β_i, and γ_i values. Logically, this update method would need to occur after each mutual authentication between a tag and a reader to maintain the low cost of not periodically updating a large amount of tags at once. However, updating the tags also presents a new problem: an attacker may still eavesdrop or tamper with the updating process. To address this issue, Juels proposed using one-time pads that have been used across multiple authentication protocols to update the current values of a tag and a reader. Using this method, a malicious third party who only eavesdrops periodically will be unlikely to gain the updated α_i, β_i, and γ_i values.

One-time pads [149] may be thought of as a simpler form of encryption than that used within cryptography; thus, they are able to be used within tags that have smaller computational capabilities (i.e., passive tags). One-time pads are essentially a random bit string of length l. If two parties then share a secret one-time pad δ, it has been proven that a message M may be sent secretly via cipher text $M \oplus \delta$ between both parties, where $\oplus$ denotes the exclusive OR (XOR) operation. Thus, after mutual authentication between both the tag and the reader, the reader uses these one-time pads to update the α_i, β_i, and γ_i values noted earlier and sends these pads to the tag so that it may update its values as well. Provided a malicious third party doesn't eavesdrop upon a reader transmitting the message and obtain these pads, they achieve no knowledge of the newly updated tag values. However, in case, a third party were to obtain one of the pads, this scheme also has an additional spin on what is considered to be the normal use of a one-time pad. This involves using the one-time encryption across multiple authentication sessions. To achieve this, pads from two different authentication sessions are XORed with a given tag value w to update it, where $w \in \alpha_i \cup \beta_i \cup \gamma_i$. Therefore, even if a third party successfully obtains a pad used in a prior session, it is seen that they will still not be able to obtain any information about the updated values of w.

The minimalist approach offers resistance against spying upon corporate businesses, such as the clandestine scanning of product stocks in a supply-chain market. Since our first developed protocol was made for retail environments, we borrowed this idea of pseudonym rotation along with Juels soft-blocking technique [145] to develop what is known as "RFIDGuard."

9.5.2 *RFIDGuard: an authentication and privacy protocol designed for passive RFID tags*

Large organizations such as Wal-Mart, Procter and Gamble, and the US Department of Defense have generated a lot of attention for RFID technology in recent years due to their need of deploying RFID as a tool for automated oversight of their supply chains [141]. This is due to a significant advantage that RFID tags have overstandard barcodes; they are able to uniquely identify individual items of a product type, unlike barcodes, which only identify each product type. However, the elongated reading range of an RFID tag enables third-party users to eavesdrop upon transmissions between authenticated readers and tags, thus enabling a security breach. RFID security within supply-chain management was our original intent for studying and creating different protocols for passive RFID tags, and thus lead to the creation of our first developed protocol, "RFIDGuard: An Authentication and Privacy Protocol Designed for Passive RFID Tags."

The idea behind RFIDGuard is comprised of a modified version of two previously developed protocols developed by Juels, the "minimalist" system [147] and the soft-blocking technique [145] as discussed earlier in the chapter. The minimalist approach suggests each tag contains an array of pseudonyms, such that upon each query from a reader, the tag will rotate through its list of pseudonyms, providing protection from any malicious third parties since they will not have the entire list themselves. To establish mutual authentication between a reader and a tag, our protocol uses this pseudonym rotation concept; however, unlike the minimalist approach, it will not need to update the tag's pseudonyms or contain additional keys to perform this task. The original soft-blocking approach interacts with the singulation protocol used to distort multiple RFID tags scanned at the same time by blocking part of the tree using what's known as a "privacy bit." Our protocol establishes privacy to RFID tags using a privacy bit, but instead of interacting with the singulation protocol, the privacy bit toggles the tags between locked and unlocked states. During a locked state, tags will only provide enough information to an authenticated reader to unlock it where it will then supply its EPC [140] to the reader, whereas an unauthorized reader would not contain enough information to retrieve the secret data. The rest of this section provides a more detailed look at our protocol and its security measures.

The RFIDGuard protocol is comprised of four smaller protocols: *in-store, checkout, out-store,* and *return.* Each protocol represents a presumed location of a tag, thus providing it with different amounts of security. Each tag contains several different items: a list of pseudonyms, a pointer representing what pseudonym the tag is currently pointing to, a number representing a generic *name* of the product, a privacy bit contained as the first bit in a tag's EPC, and any secret information a tag may need to send to a reader upon authentication including the aforementioned EPC (referred to as *ID* in this protocol). Each tag is assumed to have no further interaction when it is made until it has reached the retail environment and therefore it will start in the in-store/checkout protocol. This is denoted by a tag's privacy bit starting with 0 to denote the tag in an "unlocked" state. The in-store protocol is simply the first two steps within the checkout protocol (as is the out-store protocol the first two steps of the return protocol), so we will only refer to them as such.

Upon query of a tag within the checkout protocol, a tag will make a copy of its current pointer sending it, its *name*, and its *ID* to the tag, while increasing its pointer an additional time after transferring the message. The reason in which the pointer is increased upon every query will be explained in the following protocol. As stated earlier, the tag is assumed to have no further interaction until it has arrived to the store;

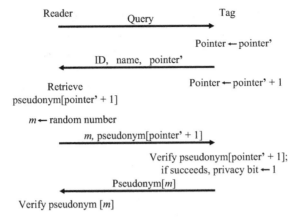

Figure 9.5 The checkout protocol of RFIDGuard

therefore, any query beyond that point while still in the unlocked state is assumed to be given by an authenticated reader and therefore will not need additional security before giving out its *ID*. Upon receiving the tag's message, the reader will use the tag's generic *name* and retrieve the pseudonym after the location of the pointer given, pseudonym[pointer+1]. Only an authenticated reader will know that you should return the pseudonym after the pointer given and not its current one, thus upon verification by the tag, the tag will flip its privacy bit to 1, denoting it to be within the "locked" state and the tag will follow the return protocol until it has been unlocked again. Before this occurs, however, the reader also sends an additional random number *m* denoting a pointer location within the tag where $1 \leq m \leq n$ for *n* pseudonyms in a tag. After verification of the reader, the tag will also send pseudonym[*m*], the pseudonym according to the location of the tag. This allows the protocol to combat against cloned tags. Some cloned tags may contain valid pseudonyms of an authentic tag but not contain all of them or have them placed in the incorrect order. If this is true, the query for a random pseudonym will have a great chance of catching this, thus if the pseudonym[*m*] isn't valid or the tag doesn't send one in a designated timing period, the reader will alarm the store that the consumer is carrying a corrupt or cloned tag. A pictorial example of this tag is given in Figure 9.5.

As earlier stated, after the tag has completed the checkout protocol, it will enter the locked state by setting its privacy bit to 1 where it will follow the return protocol procedures until it is unlocked. In this locked state, the tag is assumed to be present to malicious third-party readers and will not give information as easily as it has in the previous protocol. Therefore upon query, the tag will repeat the same process with its pointer but only its copy and its *name*, not its *ID*. The reason in which the pointer is increased upon every query is more prevalent in this subprotocol than in the previous one. Any hacker trying to repeatedly scan a tag for its information will trigger the tag to continue switch its current pointer. Therefore, if the hacker were to retrieve one or two pseudonyms from the reader within the checkout protocol, it would confuse the

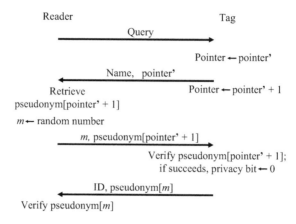

Figure 9.6   The return protocol of RFIDGuard

hacker as to which position they should be in. This would enable the reader to better catch a cloned tag made from the stolen pseudonym and not verify the tag. Returning to the subject of the *return* protocol, a valid reader would take the information sent from the tag and repeat the process of sending the tag a random pointer m and the pseudonym[pointer+1] to verify itself. Upon verification of the reader, the tag will not only send pseudonym[m] but its *ID* as well. The purpose of sending pseudonym[m] here is not to check for clones, but to check for corrupted tags. A valid reader would only be able to pick up a tag within a store for this protocol if a consumer is returning an item; thus, one may corrupt the tag to increase its original value worth, enabling the consumer to receive a larger cash amount or exchange the item for a higher priced one. Thus, this randomized pseudonym feature would try and catch corrupted tags before the consumer is able to steal money from the business. A pictorial example of this tag is given within Figure 9.6.

This protocol achieves many of the security requirements stated within [136] as listed earlier in the chapter; however, it may not be clear as to why certain items within the tag are disclosed within certain parts of the protocol. By looking at checkout protocol, one would wonder why the *ID* is released before validating the reader or why multiple pseudonyms are given on both subprotocols. The reason for both of these questions are because we make an assumption of both protocols (at least ones with authenticated readers) being executed within a physical building. Therefore, given the limited broadcasting range of a passive RFID tag, it would be very unlikely for someone to eavesdrop upon the transmission of messages during the execution of these protocols for an extended period of time. Furthermore, since our protocol is contained to a physical building, we assume that a retail store has some security mechanisms that prevent unauthorized readers from entering the store. This can be easily achieved by installing detection devices near the entrance of the store to detect unauthorized readers [161].

Another security issue that needs to be discussed within this protocol deals with the requirement of preventing a tracking attack. As stated earlier in discussing the

HashTree protocol, tracking involves knowing the whereabouts of a tag by continually receiving a static value returned from it. This issue deals with the generic name used to search for tag's information within the protocol. There is no additional device method that is needed to combat this issue, but more of an understanding of what the generic name value represents. This *name* stands for a product type and not every individual product within the store. For example, two bottles of soda may have the same generic name as long they are made by the same company. If the protocol was not setup this way, the back-end database of the system would exponentially increase in cost, both physically and memory wise, using extra slots that may never need to be returned such every individual bottle of soda.

Upon explanation of the above concerned security requirements, RFIDGuard has the potential of becoming the standard retail environment protocol with further research.

9.6   Conclusion

RFID is a promising technology that can revolutionize the way we lead our lives. However, before this becomes a reality, certain security issues like consumer privacy protection and fraud prevention and detection must be addressed. This chapter covers only a minimum amount of protocols used within RFID technology to establish the current security that protects this field. Before RFID officially overtakes the use of barcodes, extensive research must continue to ensure an efficient speed and protection of the individual tags. However, it still is believed that not every system using passive RFID tags will encase the same approach, but rather a number of security protocols will become the standard by EPC regulations for each separate RFID system. There is an ample scope in the field of RFID security to improve and innovate in order to allow RFID technology to be incorporated into our daily lives.

References

[1] http://en.wikipedia.org/wiki/Distribution_center.

[2] *The Amazon Warehouses.* http://imgur.com/gallery/uHZbW, 2013.

[3] M. Backes, T. R. Gross, and G. Karjoth. *Tag Identification System*, U.S. Patent Office, 2008.

[4] J. I. Capetanakis, "Tree Algorithms for Packet Broadcast Channels," *IEEE Transactions on Information Theory*, vol. 25, pp. 505–515, 1979.

[5] B. Carbunar, M. K. Ramanathan, M. Koyuturk, C. Hoffmann, and A. Grama, "Redundant Reader Elimination in RFID Systems," *Proc. IEEE Communications Society Conf on SECON*, pp. 576–580, 2005.

[6] R. Dorfman, "The Detection of Defective Members of Large Populations," *Annals of Mathematical Statistics*, vol. 14, pp. 436–440, 1943.

[7] EPCGlobal Inc. *EPC Radio-Frequency Identity Protocols Generation-2, Protocol for Communications at 860 MHz–960 MHz*, 2.0.0 edition, November 2013.

[8] K. Finkenzeller. *RFID Handbook: Fundamentals and Applications in Contactless Smart Cards, Radio Frequency Identification and Near-Field Communication.* Wiley, Chichester, 2010.

[9] P. Flajolet and G. N. Martin, "Probabilistic Counting Algorithms for Data Base Applications," *Journal of Computer and System Sciences*, vol. 31, no. 2, pp. 182–209, 1985.

[10] K. Fyhn, R. M. Jacobsen, P. Popovski, and T. Larsen, "Fast Capture – Recapture Approach for Mitigating the Problem of Missing RFID Tags," *IEEE Transactions on Mobile Computing*, vol. 11, no. 3, pp. 518–528, 2012.

[11] H. Han, B. Sheng, C. C. Tan, Q. Li, W. Mao, and S. Lu, "Counting RFID Tags Efficiently and Anonymously," *Proc. of IEEE INFOCOM*, 2010.

[12] R. Jacobsen, K. F. Nielsen, P. Popovski, and T. Larsen, "Reliable Identification of RFID Tags using Multiple Independent Reader Sessions," *Proc. IEEE Int. Conf. RFID*, pp. 64–71, 2009.

[13] R. K. Jain, D.-M. W. Chiu, and W. R. Hawe. A quantitative measure of fairness and discrimination for resource allocation in shared computer systems. Technical report, Digital Equipment Corporation, 1984.

[14] C. Law, K. Lee, and K.-Y. Siu, "Efficient Memoryless Protocol for Tag Identification," *Proc. 4th Int. Workshop on Discrete Algorithms and Methods for Mobile Computing and Communications*, 2000.

[15]   C. H. Lee and C.-W. Chung, "Efficient Storage Scheme and Query Processing for Supply Chain Management Using RFID," *Proc. ACM Conf. Management of Data*, pp. 291–302, 2008.

[16]   J. Myung and W. Lee, "Adaptive Splitting Protocols for RFID Tag Collision Arbitration," *Proc. 7th ACM Int. Symposium on Mobile Ad Hoc Networking and Computing*, pp. 202–213, 2006.

[17]   V. Namboodiri and L. Gao, "Energy-aware Tag Anticollision Protocols for RFID Systems," *Proc. 5th IEEE Int. Conf. Pervasive Computing and Communications*, pp. 23–36, 2007.

[18]   B. Nath, F. Reynolds, and R. Want, "RFID Technology and Applications," *Proceeding of IEEE Pervasive Computing*, vol. 5, pp. 22–24, 2006.

[19]   A. Nemmaluri, M. D. Corner, and P. Shenoy, "Sherlock: Automatically Locating Objects for Humans," *Proc. Int. Conf. Mobile Systems, Applications, and Services*, pp. 187–198, 2008.

[20]   L. M. Ni, Y. Liu, Y. C. Lau, and A. P. Patil, "LANDMARC: Indoor Location Sensing Using Active RFID," *Wireless Networks*, vol. 10, pp. 701–710, 2004.

[21]   L. Pan and H. Wu, "Smart Trend-traversal: A Low Delay and Energy Tag Arbitration Protocol for Large RFID Systems," *Proc. IEEE INFOCOM*, 2009.

[22]   C. Qian, Y. Liu, H. Ngan, and L. M. Ni, "ASAP: Scalable Identification and Counting for Contactless RFID Systems," *Proc. 30th IEEE Int. Conf. Distributed Computing Systems*, pp. 52–61, 2010.

[23]   C. Qian, H. Ngan, and Y. Liu, "Cardinality Estimation for Large-scale RFID Systems," *Proc. 6th Annual IEEE Int. Conf. Pervasive Computing and Communications*, pp. 30–39, 2008.

[24]   M. Roberti, "A 5-cent Breakthrough," *RFID Journal*, 2006.

[25]   W. A. Rosenkrantz and D. Towsley, "On the Instability of Slotted Aloha Multiaccess Algorithm," *IEEE Transactions on Automatic Control*, vol. 28, no. 10, pp. 994–996, 1983.

[26]   A. S. Tanenbaum. *Computer Networks*. Prentice-Hall, Upper Saddle River, NJ, 2002.

[27]   S. Tang, J. Yuan, X.-Y. Li, G. Chen, Y. Liu, and J. Zhao, "Raspberry: A Stable Reader Activation Scheduling Protocol in Multi-Reader RFID Systems," *Proc. IEEE Int. Conf. Network Protocols*, pp. 304–313, 2009.

[28]   J. Waldrop, D. W. Engels, and S. E. Sarma, "Colorwave: A MAC for RFID Reader Networks," *Proc. IEEE Wireless Communications and Networking*, pp. 1701–1704, 2003.

[29]   C. Wang, H. Wu, and N.-F. Tzeng, "RFID-based 3-D Positioning Schemes," *Proc. IEEE INFOCOM*, pp. 1235–1243, 2007.

[30]   B. Zhen, M. Kobayashi, and M. Shimizu, "Framed ALOHA for Multiple RFID Objects Identification," *IEICE Transactions on Communications*, vol. 88, pp. 991–999, 2005.

[31]   Z. Zhou, H. Gupta, S. R. Das, and X. Zhu, "Slotted Scheduled Tag Access in Multi-reader RFID Systems," *Proc. IEEE Int. Conf. Network Protocols*, pp. 61–70, 2007.

[32] J. Collins and S. E. Zweig, "Sensing a Product's Shelf Life," *RFID Journal*, 2005.

[33] M. Roberti, "Wal-mart Relaunches EPC RFID Effort, Starting with Men's Jeans and Basics," *RFID Journal*, 2010.

[34] C. Swedberg, "Honeywell Aerospace Tags Parts for Airbus," *RFID Journal*, 2013.

[35] C. Swedberg, "Dutch Researchers Focus on RFID-based Sensors for Monitoring Apnea, Epilepsy," *RFID Journal*, 2007.

[36] M. C. O'Connor, "HP Kicks off US RFID Demo Center," *RFID Journal*, 2004.

[37] "Sprinkling RFID Sensor Tags from The Sky," 2006, https://rfidinjapan. wordpress.com/2006/03/20/sprinkling-rfid-sensor-tags-from-the-sky/.

[38] "Tempcorder 2.4GHz Active RFID Tag Series," http://www.hk-rfid.com/home/news201211b_Tempcorder%20Case_121109.pdf.

[39] S. Ohkubo and K. Takiishi, "Technologies to Reduce Power Consumption of Active RFID Tags," *NTT DoCoMo Journal*, vol. 8, no. 1, pp. 33–40, 2006.

[40] A. Amanna, A. Agrawal, and M. Manteghi, "Active RFID for Enhanced Railway Operations," *Proc. ASME Rail Transport Division Conf.*, 2010.

[41] "TI's RFID Tagging System Tracks and Acoustic Transponders in Offshore Seismic Exploration," *Assembly Automation*, vol. 20, no. 1.

[42] "Tracking & Tracing at BMW," http://www.brooks.com/applications-by-industry/semiconductor/rfid/rfid-projects/automotive/tracking-and-tracing-at-bmw.

[43] M. C. O'Connor, "Boeing Wants Dreamliner Parts Tagged," *RFID Journal*, 2005.

[44] M. Shahzad and A. X. Liu, "Probabilistic Optimal Tree Hopping for RFID Identification," *Proc. ACM SIGMETRICS*, Pittsburgh, PA, 2013, pp. 293–304.

[45] M. Shahzad and A. X. Liu, "Every Bit Counts – Fast and Scalable RFID Estimation," *Proc. ACM MobiCom*, 2012, pp. 365–376.

[46] R. Ferrero, F. Gandino, B. Montrucchio, and M. Rebaudengo, "Fair Anti-collision Protocol in Dense RFID Networks," *Proc. of 3rd Int. EURASIP Workshop on RFID Technology*, 2010, pp. 101–105.

[47] L. Chen, I. Demirkol, and W. Heinzelman, "Token MAC: a Fair MAC Protocol for Passive RFID Systems," *Proc. IEEE GLOBECOM*, 2011.

[48] R. Sámano-Robles and A. Gameiro, "Throughput, Stability and Fairness of RFID Anti-collision Algorithms with Tag Cooperation," *The International Journal of Advances in Telecommunications*, vol. 5, pp. 229–238, 2012.

[49] H. Vogt, "Efficient Object Identification with Passive RFID Tags," *Pervasive Computing*, vol. 2414, pp. 98–113, 2002.

[50] M. Kodialam and T. Nandagopal, "Fast and Reliable Estimation Schemes in RFID Systems," *Proc. of ACM MobiCom*, 2006, pp. 322–333.

[51] C. Bordenave, D. McDonald, and A. Proutiere, "Performance of Random Medium Access Control, An Asymptotic Approach," *Proc. ACM SIGMETRICS*, 2008.

[52] J.-R. Cha and I. Jae-Hyun Kim, "Novel Anti-collision Algorithms for Fast Object Identification in RFID System," *Proc. of Int. Conf. Parallel and Distributed Systems*, 2005.

[53] EPCGlobal Inc. *Radio-Frequency Identity Protocols Class-1 Generation-2 UHF RFID Protocol for Communications at 860 MHz–960 MHz*, 1.2.0 edition, 2008.

[54] M. Kodialam, T. Nandagopal, and W. C. Lau, "Anonymous Tracking Using RFID Tags," *Proc. IEEE INFOCOM*, 2007.

[55] T. Li, S. Wu, S. Chen, and M. Yang, "Energy Efficient Algorithms for the RFID Estimation Problem," *Proc. IEEE INFOCOM*, 2010.

[56] Philips-Semiconductors. *SL2 ICS11 I.Code UID Smart Label IC Functional Specification Datasheet*, www.advanide.com/datasheets/sl2ics11.pdf, 2004.

[57] V. Shah-Mansouri and V. W. Wong, "Cardinality Estimation in RFID Systems With Multiple Readers," *Proc. IEEE GLOBECOM*, 2009.

[58] A. Zanella, "Estimating Collision Set Size in Framed Slotted Aloha Wireless Networks and RFID Systems," *IEEE Communications Letters*, vol. 16, no. 3, pp. 300–303, 2012.

[59] X. Liu, B. Xiao, K. Li, J. Wu, A. X. Liu, H. Qi, and X. Xie, "RFID Cardinality Estimation with Blocker Tags," *Proc. of IEEE INFOCOM*, 2015.

[60] L. Yang, Y. Chen, X.-Y. Li, C. Xiao, M. Li, and Y. Liu, "Tagoram: Real-Time Tracking of Mobile RFID Tags to High Precision Using COTS Devices," *Proc. of ACM MobiCom*, 2014.

[61] L. Shangguan, Z. Zhou, X. Zheng, L. Yang, Y. Liu, and J. Han, "ShopMiner: Mining Customer Shopping Behavior in Physical Clothing Stores with Passive RFIDs," *Proc. of ACM SenSys*, 2015.

[62] X. Liu, K. Li, J. Wu, *et al.*, "TOP-*k* Queries for Multi-category RFID Systems," *Proc. of IEEE INFOCOM*, 2016.

[63] J. Liu, B. Xiao, S. Chen, F. Zhu, and L. Chen, "Fast RFID Grouping Protocols," *Proc. of IEEE INFOCOM*, 2015.

[64] L. Shangguan, Z. Yang, A. X. Liu, Z. Zhou, and Y. Liu, "Relative Localization of RFID Tags using Spatial-temporal Phase Profiling," *Proc. of USENIX NSDI*, 2015.

[65] X. Liu, X. Xie, K. Li, *et al.*, "Fast Tracking the Population of Key Tags in Large-scale Anonymous RFID Systems," *IEEE/ACM Transactions on Networking*, vol. 25, no. 1, pp. 278–291, 2017.

[66] T. Liu, L. Yang, Q. Lin, and Y. Liu, "Anchor-free Backscatter Positioning for RFID Tags with High Accuracy," *Proc. of IEEE INFOCOM*, 2014.

[67] L. Xie, H. Han, Q. Li, J. Wu, and S. Lu, "Efficiently Collecting Histograms Over RFID Tags," *Proc. of IEEE INFOCOM*, 2014.

[68] J. Liu, B. Xiao, K. Bu, and L. Chen, "Efficient Distributed Query Processing in Large RFID-enabled Supply Chains," *Proc. of IEEE INFOCOM*, 2014.

[69] "http://www.centreforaviation.com/news/share-market/2010/06/17/hong-kong-airport-sets-new-cargo-traffic-record-fedex-sees-surging-asian-exports/page1."

[70]   M. Shahzad and A. X. Liu, "Fast and Accurate Estimation of RFID Tags," *IEEE/ACM Transactions on Networking*, vol. 23, no. 1, pp. 241–254, 2015.

[71]   M. Chen and S. Chen, "ETAP: Enable Lightweight Anonymous RFID Authentication with O(1) Overhead," *Proc. of IEEE ICNP*, 2015.

[72]   C. Qian, H. Ngan, Y. Liu, and L. M. Ni, "Cardinality Estimation for Large-scale RFID Systems," *IEEE Transactions on Parallel and Distributed Systems*, vol. 22, no. 9, pp. 1441–1454, 2011.

[73]   B. Chen, Z. Zhou, and H. Yu, "Understanding RFID Counting Protocols," *Proc. of ACM MobiCom*, 2013.

[74]   Y. Zheng and M. Li, "ZOE: Fast Cardinality Estimation for Large-Scale RFID Systems," *Proc. of IEEE INFOCOM*, 2013.

[75]   T. Li, S. Wu, S. Chen, and M. Yang, "Generalized Energy-Efficient Algorithms for the RFID Estimation Problem," *IEEE/ACM Transactions on Networking*, vol. 20, no. 6, pp. 1978–1990, 2012.

[76]   X. Liu, K. Li, A. X. Liu, *et al.*, "Multi-category RFID Estimation," *IEEE/ACM Transactions on Networking*, vol. 25, no. 1, pp. 264–277, 2017.

[77]   S. Lee, S. Joo, and C. Lee, "An Enhanced Dynamic Framed Slotted ALOHA Algorithm for RFID Tag Identification," *Proc. of IEEE MobiQuitous*, 2005.

[78]   M. Shahzad and A. X. Liu, "Probabilistic Optimal Tree Hopping for RFID Identification," *IEEE/ACM Transactions on Networking*, vol. 23, no. 3, pp. 796–809, 2015.

[79]   P. Semiconductors, "I-CODE Smart Label RFID Tags," http://www.nxp.com/acrobat_download/other/identification/SL092030.pdf, 2004.

[80]   M. Chen, W. Luo, Z. Mo, S. Chen, and Y. Fang, "An Efficient Tag Search Protocol in Large-Scale RFID Systems With Noisy Channel," *IEEE/ACM Transactions on Networking*, vol. 24, no. 2, pp. 703–716, 2016.

[81]   Y. Zheng and M. Li, "Fast Tag Searching Protocol for Large-Scale RFID Systems," *IEEE/ACM Transactions on Networking*, vol. 21, no. 3, pp. 924–934, 2013.

[82]   Q. Xiao, B. Xiao, and S. Chen, "Differential Estimation in Dynamic RFID Systems," *Proc. of IEEE INFOCOM*, 2013.

[83]   W. Luo, S. Chen, T. Li, and Y. Qiao, "Probabilistic Missing-tag Detection and Energy-Time Tradeoff in Large-scale RFID Systems," *Proc. of ACM MobiHoc*, pp. 95–104, 2012.

[84]   W. Gong, K. Liu, X. Miao, Q. Ma, Z. Yang, and Y. Liu, "Informative Counting: Fine-grained Batch Authentication for Large-Scale RFID Systems," *Proc. of ACM MobiHoc*, 2013.

[85]   D. E. Smith, *A Source Book in Mathematics*. Courier Dover Publications, New York, NY, 2012.

[86]   M. Schilling, "Understanding Probability: Chance Rules in Everyday Life," *The American Statistician*, vol. 60, no. 1, pp. 97–98, 2006.

[87]   http://pan.baidu.com/s/1i3YgJGh.

[88]   Y. Qiao, S. Chen, T. Li, and S. Chen, "Energy-efficient Polling Protocols in RFID Systems," *Proc. of ACM Mobihoc*, 2011.

[89] S. N. V and D.-B. IV, "Mathematische Statistik in der Technik," *Deutscher Verl. der Wissenschaften*, 1963.

[90] "Preliminary national retail security survey findings," 2012, https://nrf.com/news/national-retail-security-survey-retail-shrinkage-totaled-345-billion-2011.

[91] K. Bu, J. Bao, M. Weng, *et al.*, "Who Stole My Cheese?: Verifying Intactness of Anonymous RFID Systems," *Ad Hoc Networks*, vol. 36, pp. 111–126, 2016.

[92] B. Chen, Z. Zhou, and H. Yu, "Understanding RFID Counting Protocols," *Proceedings of the 19th Annual International Conference on Mobile Computing & Networking*, pp. 291–302. ACM, 2013.

[93] M. S. Humphreys, "Shoplifting – In 100 Words," *The British Journal of Psychiatry*, vol. 202, no. 2, pp. 128, 2013.

[94] T. Li, S. Chen, and Y. Ling, "Identifying the Missing Tags in a Large RFID System," *Proc. of ACM MobiHoc*, pp. 1–10, 2010.

[95] X. Liu, K. Li, G. Min, Y. Shen, A. Liu, and W. Qu, "Completely Pinpointing the Missing RFID Tags in a Time-efficient Way," *IEEE Transactions on Computers*, vol. 64, pp. 1–11, 2013.

[96] A. D. Smith, A. A. Smith, and D. L. Baker, "Inventory Management Shrinkage and Employee Anti-theft Approaches," *International Journal of Electronic Finance*, vol. 5, no. 3, pp. 209–234, 2011.

[97] C. C. Tan, B. Sheng, and Q. Li, "How to Monitor for Missing RFID Tags," *Proc. IEEE ICDCS*, pp. 295–302, 2008.

[98] R. Zhang, Y. Liu, Y. Zhang, and J. Sun, "Fast Identification of the Missing Tags in a Large RFID System," *Proc. IEEE SECON*, 2011.

[99] Y. Zheng and M. Li, "P-MTI: Physical-layer Missing Tag Identification via Compressive Sensing," *2013 Proceedings of IEEE INFOCOM*, pp. 917–925. IEEE, 2013.

[100] X. Liu, H. Qi, K. Li, Y. Shen, A. X. Liu, and W. Qu, "Time- and Energy-efficient Detection of Unknown Tags in Large-scale RFID Systems," *Proc. of IEEE MASS*, 2013.

[101] A. Matic, A. Papliatseyeu, V. Osmani, and O. Mayora-Ibarra, "Tuning to Your Position: FM Radio Based Indoor Localization with Spontaneous Recalibration," *Proc. of IEEE PerCom*, 2010.

[102] W. Zhu, J. Cao, Y. Xu, L. Yang, and J. Kong, "Fault-Tolerant RFID Reader Localization Based on Passive RFID Tags," *Proc. of IEEE INFOCOM*, 2012.

[103] M. Chen, W. Luo, Z. Mo, S. Chen, and Y. Fang, "An Efficient Tag Search Protocol in Large-Scale RFID Systems," *Proc. of IEEE INFOCOM*, 2013.

[104] K. Bu, B. Xiao, Q. Xiao, and S. Chen, "Efficient Misplaced-Tag Pinpointing in Large RFID Systems," *IEEE Transactions on Parallel and Distributed Systems*, vol. 23, no. 11, pp. 2094–2106, 2012.

[105] S. Qi, Y. Zheng, M. Li, L. Lu, and Y. Liu, "COLLECTOR: A Secure RFID-Enabled Batch Recall Protocol," *Proc. of IEEE INFOCOM*, 2014.

[106]   Y. Zheng and M. Li, "Fast Tag Searching Protocol for Large-Scale RFID," *Proc. of IEEE ICNP*, 2011.

[107]   H. Liu, W. Gong, L. Chen, W. He, K. Liu, and Y. Liu, "Generic Composite Counting in RFID Systems," *Proc. of IEEE ICDCS*, 2014.

[108]   B. Sheng, C. C. Tan, Q. Li, and W. Mao, "Finding Popular Categories for RFID Tags," *Proc. of ACM MobiHoc*, 2008.

[109]   F. C. Schoute, "Dynamic Frame Length ALOHA," *IEEE Transactions on Communications*, vol. 31, no. 4, pp. 565–568, 1983.

[110]   P. Popovski, K. Fyhn, R. M. Jacobsen, and T. Larsen, "Robust Statistical Methods for Detection of Missing RFID Tags," *IEEE Wireless Communications*, vol. 18, no. 4, pp. 74–80, 2011.

[111]   W. Bishop, "Documenting the Value of Merchandising," Technical Report, *National Association for Retail Merchandising Service*, Plover, WI, 2003.

[112]   Y. Yin, L. Xie, S. Lu, and D. Chen, "Efficient Protocols for Rule Checking in RFID Systems," *Proc. of IEEE ICCCN*, 2013.

[113]   L. Yang, J. Han, Y. Qi, and Y. Liu, "Identification-Free Batch Authentication for RFID Tags," *Proc. of IEEE ICNP*, 2010.

[114]   G. Bianchi, "Revisiting an RFID Identification-free Batch Authentication Approach," *IEEE Communications Letters*, vol. 15, no. 6, pp. 632–634, 2011.

[115]   C. C. Tan, B. Sheng, and Q. Li, "Efficient Techniques for Monitoring Missing RFID Tags," *IEEE Transactions on Wireless Communications*, vol. 9, no. 6, pp. 1882–1889, 2010.

[116]   W. Luo, S. Chen, T. Li, and S. Chen, "Efficient Missing Tag Detection in RFID Systems," *Proc. of IEEE INFOCOM*, 2011.

[117]   X. Liu, K. Li, G. Min, Y. Shen, A. X. Liu, and W. Qu, "Completely Pinpointing the Missing RFID Tags in a Time-Efficient Way," *IEEE Transactions on Computers*, vol. 64, no. 1, pp. 87–96, 2015.

[118]   X. Liu, K. Li, G. Min, *et al.*, "Efficient Unknown Tag Identification Protocols in Large-Scale RFID Systems," *IEEE Transactions on Parallel Distributed Systems*, vol. 25, no. 12, pp. 3145–3155, 2014.

[119]   B. Sheng, Q. Li, and W. Mao, "Efficient Continuous Scanning in RFID Systems," *Proc. of IEEE INFOCOM*, 2010.

[120]   X. Liu, S. Zhang, K. Bu, and B. Xiao, "Complete and Fast Unknown Tag Identification in Large RFID Systems," *Proc. of IEEE MASS*, 2012.

[121]   X. Liu, K. Li, Y. Shen, G. Min, B. Xiao, W. Qu, and H. Li, "A Fast Approach to Unknown Tag Identification in Large Scale RFID Systems," *Proc. of IEEE ICCCN*, 2013.

[122]   B. Bloom, "Space/time Tradeoffs in Hash Coding with Allowable Errors," *ACM Communications*, vol. 13, no. 7, pp. 422–426, 1970.

[123]   H. Fang, K. Murali, and L. TV, "Building High Accuracy Bloom Filters Using Partitioned Hashing," *ACM SIGMETRICS Performance Evaluation Review*, vol. 35, no. 1, pp. 277–288, 2007.

[124]   K. Huang, J. Zhang, D. Zhang, *et al.*, "A Multi-Partitioning Approach to Building Fast and Accurate Counting Bloom Filters," *Proc. of IEEE IPDPS*, 2013.

[125]   I. Chlamtac, C. Petrioli, and J. Redi, "Energy-Conserving Access Protocols for Identification Networks," *IEEE/ACM Transactions on Networking*, vol. 7, no. 1, pp. 51–59, 1999.

[126]   X. Liu, H. Qi, K. Li, J. Wu, W. Xue, G. Min, and B. Xiao:, "Efficient Detection of Cloned Attacks for Large-Scale RFID Systems," *Proc. of Springer ICA3PP*, 2014.

[127]   Y. Qiao, T. Li, and S. Chen, "One Memory Access Bloom Filters and Their Generalization," *Proc. of IEEE INFOCOM*, 2011.

[128]   S. Zhang, X. He, H. Song, and D. Zhang, "Time efficient tag searching in multiple reader RFID systems," in *Proc. Green Computing and Communications (GreenCom)*, IEEE, 2013.

[129]   X. Liu, K. Li, H. Qi, B. Xiao, and X. Xie, "Fast Counting the Key Tags in Anonymous RFID Systems," *Proc. of IEEE ICNP*, 2014.

[130]   L. Yang, P. Pai, F. Dang, C. Wang, X.-Y. Li, and Y. Liu, "Anti-counterfeiting via Federated RFID Tags' Fingerprints and Geometric Relationships," *Proc. of IEEE INFOCOM*, 2015.

[131]   Y. Zheng, M. Li, and C. Qian, "PET: Probabilistic Estimating Tree for Large-Scale RFID Estimation," *Proc. of IEEE ICDCS*, 2011.

[132]   Y. Hou, J. Ou, Y. Zheng, and M. Li, "PLACE: Physical Layer Cardinality Estimation for Large-Scale RFID Systems," *Proc. of IEEE INFOCOM*, 2015.

[133]   L. Kong, L. He, Y. Gu, M.-Y. Wu, and T. He, "A Parallel Identification Protocol for RFID Systems," *Proc. of IEEE INFOCOM*, 2014.

[134]   O. N. Maire, "Low-cost SHA-1 Hash Function Architecture for RFID Tags," *RFIDSec*, vol. 8, pp. 41–51, 2008.

[135]   L. Yang, J. Han, Y. Qi, C. Wang, T. Gux, and Y. Liu, "Season: Shelving Interference and Joint Identification in Large-Scale RFID Systems," *Proc. of IEEE INFOCOM*, 2011.

[136]   T. Dimitriou, "A Secure and Efficient RFID Protocol That Could Make Big Brother (Partially) Obsolete," *PERCOM'06: Proceedings of the Fourth Annual IEEE International Conference on Pervasive Computing and Communications*, pp. 269–275, Washington, DC, USA, 2006. IEEE Computer Society.

[137]   D. Molnar, A. Soppera, and D. Wagner, "A Scalable, Delegatable Pseudonym Protocol Enabling Ownership Transfer of RFID Tags," Cryptology ePrint Archive, Report 2005/315, 2005. http://eprint.iacr.org/.

[138]   D. Molnar and D. Wagner, "Privacy and Security in Library RFID: Issues, Practices, and Architectures," *CCS'04: Proceedings of the 11th ACM Conference on Computer and Communications Security*, pp. 210–219, New York, NY, USA, 2004. ACM.

[139]   D. Johnson, C. Perkins and J. Arkko, "Mobility Support in IPv6," RFC 3775, IETF, June 2004.

[140]   EPCglobal. Epcglobal website. http://www.EPCglobalinc.org/, 2007.

[141]   A. Juels, "RFID Security and Privacy: A Research Survey," *IEEE Journals on Selected Areas in Communications*, vol. 24, no. 2, pp. 381–394, 2006.

[142]   S.E. Sarma, S.A. Weis, D.W. Engels, "RFID Systems and Security and Privacy Implications," Workshop on Cryptographic Hardware and Embedded Systems (CHES) 2002, LNCS no. 2523, 2003, 454–469.

[143]   International Civil Aviation Organization ICAO, "Document 9303, Machine readable travel documents (MRTD), Part I," Machine readable passports, 2005.

[144]   A. Juels, R. L. Rivest, and M. Szydlo, "The Blocker Tag: Selective Blocking of RFID Tags for Consumer Privacy," *Proceedings of the 10th ACM Conference on Computer and Communication Security*, pp. 103–111, 2003.

[145]   A. Juels and J. Brainard, "Soft Blocking: Flexible Blocker Tags on the Cheap," *Proceedings of the 2004 ACM Workshop on Privacy in the Electronic Society*, pp. 1–7, 2004.

[146]   L. Lu, J. Han, L. Hu, Y. Liu, and L. M. Ni, "Dynamic Key-updating: Privacy-preserving Authentication for RFID Systems," *PERCOM'07: Proceedings of the Fifth IEEE International Conference on Pervasive Computing and Communications*, pp. 13–22, Washington, DC, USA, 2007. IEEE Computer Society.

[147]   A. Juels, "Minimalist Cryptography for Low-cost RFID Tags," *Proc. 4th Int. Conf. Security Commun. Netw.*, 3352:149–164, 2004.

[148]   S.E. Sarma, "Towards the Five-cent Tag," Technical Report MIT-AUTOID-WH-006, Auto-ID Labs, 2001. http://www.autoidlabs.org/.

[149]   A.J. Menezes, P.C. van Oorschot, and S.A. Vanstone, *Handbook of Applied Cryptography*, CRC Press, Boca Raton, FL, 1996.

[150]   S. Stern, "Security Trumps Privacy," *Christian Science Monitor*, 2001.

[151]   D. Molnar and D. Wagner, "Privacy and Security in Library RFID: Issues, Practices, and Architectures," In B. Pfitzmann and P. McDaniel, editors, Proc. ACM Conf. Commun. Comput. Security, pp. 210–219, 2004.

[152]   A. Juels, D. Molnar, and D. Wagner, "Security and Privacy Issues in E-passports", *Proceedings of the First International Conference on Security and Privacy for Emerging Areas in Communications Networks (SecureComm)*, pp. 74–88, September 2005.

[153]   M. Ohkubo, K. Suzuki, and S. Kinoshita, "Cryptographic Approach to 'Privacy-friendly' Tags," *RFID Privacy Workshop*, MIT, MA, USA, November 2003.

[154]   S. Bono, M. Green, A. Stubblefield, A. Juels, A. Rubin, and M. Szydlo, "Security Analysis of a Cryptographically-enabled RFID Device," *USENIX Security Symposium*, pp. 1–16, Baltimore, Maryland, USA, July–August 2005, USENIX.

[155]   J. Wolkerstorfer, "Is Elliptic-curve Cryptography Suitable to Secure RFID Tags?," *Handout of the Encrypt Workshop on RFID and Lightweight Crypto*, July 2005.

[156]   G. Barber, E. Tsibertzopoulos, and H. B.A., "An Analysis of Using Epcglobal Class-1 Generation-2 RFID Technology for Wireless Asset Management," *Military Communications Conference*, volume 1, pp. 245–251, October 2005.

[157]   National Institute of Standards and Technology, "Secure Hash Standard," Federal Information Processing Standards Publications (FIPS PUBS), April 1995.

[158]   R. Rivest. *The MD5 Message-Digest Algorithm*. MIT Laboratory for Computer Science and RSA Data Security, Inc., Cambridge, MA, April 1992.

[159]   I. Mironov. *Hash Functions: Theory, Attacks, and Applications*. Microsoft Research, Silicon Valley Campus, Mountain View, CA, November 2005.

[160]   M. Rieback, B. Crispo, and A. Tanenbaum, "RFID Guardian: A Battery-powered Mobile device for RFID Privacy Management," *Proc. Australasian Conf. Inf. Security and Privacy*, C. Boyd and J. M. Gonzlez Nieto, Eds. New York: Springer-Verlag, 2005, vol. 3574, Lecture Notes in Computer Science, pp. 184–194.

[161]   T. Li and R. Deng, "Vulnerability Analysis of EMAP—An Efficient RFID Mutual Authentication Protocol," *International Conference on Availability, Reliability and Security*, 2007.

Index